Beginning SQL Queries

From Novice to Professional

Second Edition

Clare Churcher

Apress®

Beginning SQL Queries

Clare Churcher
Great Neck
New York, USA

ISBN-13 (pbk): 978-1-4842-1954-6   ISBN-13 (electronic): 978-1-4842-1955-3
DOI 10.1007/978-1-4842-1955-3

Library of Congress Control Number: 2016944320

Managing Director: Welmoed Spahr
Lead Editor: Jonathan Gennick
Technical Reviewer: George Anderson
Editorial Board: Steve Anglin, Pramila Balen, Louise Corrigan, Jonathan Gennick, Robert Hutchinson, Celestin Suresh John, Nikhil Karkal, James Markham, Susan McDermott, Matthew Moodie, Ben Renow-Clarke, Gwenan Spearing
Coordinating Editor: Jill Balzano
Copy Editor: April Rondeau
Compositor: SPi Global
Indexer: SPi Global
Artist: SPi Global

Distributed to the book trade worldwide by Springer Science+Business Media New York, 233 Spring Street, 6th Floor, New York, NY 10013. Phone 1-800-SPRINGER, fax (201) 348-4505, e-mail `orders-ny@springer-sbm.com`, or visit `www.springer.com`. Apress Media, LLC is a California LLC and the sole member (owner) is Springer Science + Business Media Finance Inc (SSBM Finance Inc). SSBM Finance Inc is a Delaware corporation.

For information on translations, please e-mail `rights@apress.com`, or visit `www.apress.com`.

Apress and friends of ED books may be purchased in bulk for academic, corporate, or promotional use. eBook versions and licenses are also available for most titles. For more information, reference our Special Bulk Sales–eBook Licensing web page at `www.apress.com/bulk-sales`.

Any source code or other supplementary material referenced by the author in this text is available to readers at `www.apress.com`. For detailed information about how to locate your book's source code, go to `www.apress.com/source-code/`.

Printed on acid-free paper

To Mark and Ali

Contents at a Glance

Contents

About the Author

Clare Churcher was a senior academic at Lincoln University, Christchurch, New Zealand, for twenty years and won a teaching award for her contribution to developing and delivering several undergraduate and postgraduate courses, including the analysis and design of databases. Following her time at Lincoln, she spent two years as a business analyst at Orion Health Software. She is currently developing graduate-level software courses for Tai Poutini Polytechnic, Christchurch, New Zealand.

About the Technical Reviewer

George Anderson Jr. is a working database administrator with nearly a decade of SQL experience. He credits his exposure to SQL, both at work and through the SQL community, with providing him many great opportunities for learning, growing, and networking. When not protecting data and writing code, George enjoys reading, playing golf very poorly, and spending time with his family.

Acknowledgments

First of all, many, many thanks to my husband, Neville, for reading every chapter and providing so many valuable suggestions. I would like to acknowledge one of my readers, Scott Lawley, who has given me helpful feedback and suggested the terms *process approach* and *outcome approach* as being friendlier than *algebra* and *calculus*. Thank you to my editor, Jonathan Gennick, for making this second edition possible, and to Jill Balzano for her excellent coordination. Thanks also to my employer, Tai Poutini Polytechnic, for its support.

Introduction

Overview

The syntax of SQL is quite easy to learn. A few basic ideas and a handful of keywords allow you to tackle a huge range of queries. However, many users often find themselves completely stumped when faced with a particular problem. It isn't really a great deal of help for someone to say "this is how I would do it." What you need is a variety of ways to get started on a tricky problem. Once you have made a start on a query, you need to be able to check, amend, and refine your solution until you have what you need.

Two-Pronged Approach

Throughout the book I have approached different types of queries from two directions. The two approaches have their roots in the formal relational algebra and calculus. In the body of the book I have kept the descriptions non-mathematical, however, Appendix 2 provides an introduction to the formal notation for those keen to understand the underlying theory. The first approach, which I've called the *process approach*, looks at *how* tables need to be manipulated in order to retrieve the subset of data required. You will find explanations of the different types of operations that can be performed on tables; e.g., joins, intersections, selections. Explanations are provided to help you decide which of these might be useful in particular situations. Once you understand what operations are needed, translating them into SQL is relatively straightforward.

The second approach is what I use when I just can't figure out what operations will give me the required results. This approach, which I've called the *outcome approach*, lets you describe what an expected row in your result might be like – i.e., what conditions it must obey. By looking at the data, it is surprisingly easy to develop a semi-formal description of what a "correct" retrieved row would be like (and by implication, how you would recognize an "incorrect" row). Translating this semi-formal description into a working query is straightforward.

I am always surprised at which approach my students take when confronting a new problem. Some will instantly see the operations that are needed, and others will find the outcome approach more intuitive. The choice of approach changes from query to query, from person to person, and (I suspect) from day to day. Having more than one way to get started means you are less likely to get completely baffled by a new problem.

Who Is This Book For?

This book is for anyone who has a well-designed relational database and needs to extract information from it. You might have noticed in the previous sentence that the database must be "well designed." I can't overemphasize this point. If your database is badly designed, then it will not be able to store accurate and consistent data, and so the information your queries retrieve will always be prone to inaccuracies. If you are looking to design a database from scratch, you should read my first book *Beginning Database Design*.[1] The final chapter in this book will outline a few common design problems you are likely to come across and give some advice about how to mitigate the impact or correct the problem.

Objective of This Book

In this book you will be introduced to all the main techniques and keywords needed to create SQL queries. You will learn about joins, intersections, unions, differences, selection of rows, and projection of columns. You will see how to implement these ideas in different ways using simple and nested queries, and you will be introduced to a variety of techniques for aggregating and summarizing data, including the use of window functions. You will also learn how you can investigate and improve the efficiency of your queries.

Most important of all, you will learn different ways to get started on a troublesome problem. There are almost always several different ways to express a query, and my objective is that for any particular situation I will provide you with a method of attack that matches your psyche and mood (just kidding).

New in the Second Edition

I have added a chapter on window functions describing the functionality these recently introduced concepts give to aggregating and summarizing data.

An appendix that provides an easily understood introduction to formal relational concepts and notation is also included.

---

[1]Clare Churcher, *Beginning Database Design: From Novice to Professional* (New York: Apress, 2012).

CHAPTER 1

Relational Database Overview

SQL (Structured Query Language) enables us to create tables, apply constraints, and manipulate data in a database. In this book we will concentrate on queries that allow us to extract information from a database by describing the subset of data we need. That data might be a single number, such as a product price, a list of the names of members with overdue subscriptions, or a calculation, such as the total dollar amount of products sold in the past 12 months. In this book we will be looking at different ways to approach a query so that it can be expressed correctly in SQL.

Before getting into the nuts and bolts of how to specify queries, we will review some of the ideas and terminology associated with relational databases. We will also look at data models, which are a succinct way of depicting how a particular database is put together, that is, what data is being kept where and how everything is interrelated.

It is imperative that the underlying database has been designed to accurately represent the situation it is dealing with. This means not only that suitable tables have been created, but also that appropriate constraints have been applied so that the data is consistent and stays consistent as the database evolves. Even with all the fanciest SQL in the world, you are unlikely to get accurate responses to queries if the underlying database design is faulty. If you are setting up a new database, you should refer to a design book[1] before embarking on the project.

Introducing Database Tables

In simple terms, a relational database is a set of *tables*.[2] Each table in a well-designed database keeps information about aspects of one thing, such as customers, sales, teams, or tournaments. Throughout the book we will base the majority of the examples on a database for a golf club. The tables will be introduced as we progress, and an overview is provided in Appendix 1.

Electronic supplementary material The online version of this chapter (doi:10.1007/978-1-4842-1955-3_1) contains supplementary material, which is available to authorized users.

---

[1]For example, you can refer to my other Apress book, *Beginning Database Design: From Novice to Professional* (New York: Apress, 2012).
[2]More correctly, it's a set of relations. In the body of the book common words such as *table* and *row* are used. In Appendix 2 we introduce the more formal vocabulary and notation.

© Clare Churcher 2016
C. Churcher, *Beginning SQL Queries*, DOI 10.1007/978-1-4842-1955-3_1

Attributes

When a table is created we need to specify what information it will hold. For example, a Member table might contain information about names, addresses, and contact details. We need to decide what the individual pieces of data will be. For example, we might choose to separate the name information into a title, a first name, a family name, initials, and a preferred name. This type of separation allows us more flexibility in how the data is used. For example, we can address correspondence to Mr. J. A. Stevens and start the message with *Dear Jim*. Each of these separate pieces of information is an *attribute* of the table.

To define an attribute we need to provide a name (e.g., FamilyName, Handicap, or DateOfBirth) and a domain or type. A *domain* is a set of allowed values and might be something very general or something quite specific. For example, the domain for columns storing dates might be any valid date (so that *February 29* is allowed only in leap years), whereas for columns keeping quantities the domain might be integer values greater than 0. We might initially think that the domain for a FamilyName attribute could be any string of characters, but on reflection we will need to consider whether some punctuation is allowed (probably yes), if numbers are permitted (hard to say), and if there should be a minimum or maximum length. All database systems have built-in domains or *types* such as text, integer, or date that can be chosen for each of the fields in a table. More sophisticated products allow the user to define their own types, which can be used across tables. For example, we might define a type called CarRegistration that has a predetermined template of letters and digits. Even if it is not possible to define your own types, all good database systems allow the designer to specify constraints on a particular attribute in a table. For example, in a particular table we might specify that a birthdate is a date in the past or that a handicap is between 0 and 40. Some attributes might be allowed to be empty, while others may be required to have a value.

When we view the table, the names of the attributes are the column headers, and the domain or type provides the set of allowed values. Once we have defined the table we add data by providing a row for each instance. For example, if we have a Member table, as in Figure 1-1, each row represents one member.

MemberID	LastName	FirstName	Handicap	JoinDate	Gender
118	McKenzie	Melissa	30	28-May-05	F
138	Stone	Michael	30	31-May-09	M
153	Nolan	Brenda	11	12-Aug-06	F
176	Branch	Helen		06-Dec-11	F
178	Beck	Sarah		24-Jan-10	F
228	Burton	Sandra	26	09-Jul-13	F
235	Cooper	William	14	05-Mar-08	M
239	Spence	Thomas	10	22-Jun-06	M
258	Olson	Barbara	16	29-Jul-13	F
286	Pollard	Robert	19	13-Aug-13	M
290	Sexton	Thomas	26	28-Jul-08	M
323	Wilcox	Daniel	3	18-May-09	M
331	Schmidt	Thomas	25	07-Apr-09	M
332	Bridges	Deborah	12	23-Mar-07	F
339	Young	Betty	21	17-Apr-09	F
414	Gilmore	Jane	5	30-May-07	F
415	Taylor	William	7	27-Nov-07	M
461	Reed	Robert	3	05-Aug-05	M
469	Willis	Carolyn	29	14-Jan-11	F
487	Kent	Susan		07-Oct-10	F

Figure 1-1. *The Member table*

The Primary Key

One of the most important features of a relational database table is that each of its rows should be unique. No two rows in a table should have identical values for every attribute. If we consider our member data, it is clear why this uniqueness constraint is so important. If, in the table in Figure 1-1, we had two identical rows (say, for Brenda Nolan), we would have no way to differentiate them. We might associate a team with one row and a subscription payment with the other, thereby generating all sorts of confusion.

The way that a relational database maintains the uniqueness of rows in a table is by specifying a primary key. A *primary key* is an attribute, or set of attributes, that is guaranteed to be different in every row of a given table. For data such as the member data in this example, we cannot guarantee that all our members will have different names or addresses (a father and son may share a name and address and both belong to the club). It is important that there are sufficient attributes to be able to distinguish the rows in a table. Adding a birthdate would resolve the problem mentioned above. Dealing with large numbers of attributes as a primary key can become cumbersome, so to help distinguish different members, we have included an ID number as one of the attributes in the table in Figure 1-1. We can now uniquely identify a member by specifying their ID. This has the added advantage that we can also keep track of members if they change their names. Adding an identifying number (sometimes referred to as a *surrogate key*) is very common in database tables. If MemberID is defined as the primary key for the Member table, then the database system will ensure that in every row the value of MemberID is different. The system will also ensure that the primary key field always has a value. That is, we can never add a row that has an empty MemberID field. These two requirements for a primary key field (uniqueness and not being empty) ensure that given a value for MemberID, we can always find a single row that represents that member. We will see that this is also important when we start looking at relationships between tables later in this chapter.

The code that follows shows the SQL code for creating the Member table shown in Figure 1-1. Each attribute has a name and type specified. In SQL, the keyword INT means an integer or non-fractional number, and CHAR(n) means a string of characters n long. The code also specifies that MemberID will be the primary key. Every table in a well-designed database should have a primary key clause.

```
CREATE TABLE Member (
MemberID INT PRIMARY KEY,
LastName CHAR(20),
FirstName CHAR(20),
Handicap INT,
JoinDate DATETIME,
Gender CHAR(1));
```

Inserting and Updating Rows in a Table

The emphasis of this book is on getting accurate information out of a database, but the data first has to get in somehow. Most database application developers will provide user-friendly interfaces for inserting data into the various tables. Often a form is presented to the user for entering data that may end up in several tables. Figure 1-2 shows a simple Microsoft© Access form that allows a user to enter and amend data in the Member table.

Member

MemberID	118
LastName	McKenzie
FirstName	Melissa
Handicap	30
JoinDate	28-May-05
Gender	F

Figure 1-2. *A form allowing entry and updating of data in the Member table*

It is possible to construct web forms or use mechanical readers, such as bar-code readers, that can collect data and insert it into a database. Data can also be added with bulk updates from files or be imported from other applications. Behind all the different mechanisms for updating data, SQL update queries are generated. We will see three types of queries for inserting or changing data just to get an idea of what they look like.

The code that follows shows the SQL to enter one complete row in our Member table. The data items are in the same order as specified when the table was created. Note that the date and string values need to be enclosed in single quotes.

```
INSERT INTO Member
VALUES (118, 'McKenzie', 'Melissa', '963270', 30, '05/10/1999', 'F')
```

If many of the data items are empty, we can specify which attributes will have values. If we had only the ID and last name of a member, we could insert just those two values as shown here:

```
INSERT INTO Member (MemberID, LastName)
VALUES (258, 'Olson')
```

When adding a new row as just seen, we always have to provide a value for the primary key.

We can also alter records that are already in the database with an update query. The following query will find the row for the member with ID 118 and then will update the phone number:

```
UPDATE Member
SET Phone = '875077'
WHERE MemberID = 118
```

This query specifies which rows are to be changed (the WHERE clause) and also specifies the field to be updated (the SET clause).

Designing Appropriate Tables

Even a quite modest database system will have hundreds of attributes: names, dates, addresses, quantities, prices, descriptions, ID numbers, and so on. These all have to find their way into tables, and getting them in the right tables is critical to the overall accuracy and usefulness of the database. Many problems can arise from having attributes in the "wrong" tables. As a simple illustration of what can go wrong, I'll briefly show the problems associated with having redundant information.

Say we want to add teams and practice nights to the information we are keeping about members of our golf club. We could add these two fields to the Member table, as in Figure 1-3.

MemberID ⌄	LastName ⌄	FirstName ⌄	Team ⌄	PracticeNight ⌄
286	Pollard	Robert	TeamB	Tuesday
339	Young	Betty	TeamB	Tuesday
153	Nolan	Brenda	TeamB	Monday
235	Cooper	William	TeamB	Tuesday
461	Reed	Robert	TeamA	Monday
415	Taylor	William	TeamA	Monday
414	Gilmore	Jane	TeamA	Monday
323	Wilcox	Daniel	TeamA	Monday
138	Stone	Michael		
176	Branch	Helen		

Figure 1-3. *Possible Member table*

Immediately, we can see there has been a problem with the data entry because Brenda Nolan has a practice night that is different from the rest of her team members. The piece of information about the practice night for each team is being stored several times, so inevitably inconsistencies will arise. If we formulated a query to find the practice night for TeamB, what would we expect for an answer? Should it be Monday, Tuesday, or both?

The problem here is that (in database parlance) the table is not properly *normalized*. Normalization is a formal way of checking whether attributes are in the correct table. It is outside the scope of this book to delve into normalization, but I'll just briefly show you how to avoid the problem in this particular case.

The problem is that we are trying to keep information about two different things in our Member table: information about each member (IDs, names, and so on) and information about teams (the practice nights). The PracticeNight attribute is in the wrong table. Figure 1-4 shows a better solution with two tables: one for information about members and one for information about teams.

MemberID ▾	LastName ▾	FirstName ▾	Team ▾
286	Pollard	Robert	TeamB
339	Young	Betty	TeamB
153	Nolan	Brenda	TeamB
235	Cooper	William	TeamB
461	Reed	Robert	TeamA
415	Taylor	William	TeamA
414	Gilmore	Jane	TeamA
323	Wilcox	Daniel	TeamA
138	Stone	Michael	
176	Branch	Helen	

TeamName ▾	PracticeNight ▾
TeamA	Tuesday
TeamB	Monday

Member Table   Team Table

Figure 1-4. *Member and Team tables*

This separation of information into two tables prevents the inconsistent data we had previously. The practice night for each team is stored only once. If we need to find out what night Brenda Nolan should be at practice, we now need to consult two tables: the Member table to find her team and then the Team table to find the practice night for that team. The bulk of this book is about how to do just that sort of data retrieval.

Introducing Data Models

Even the simplest databases are likely to have several tables. A *data model* is a conceptual model of the underlying data and how it is interrelated. We will use the class diagram notation from the Unified Modeling Language (UML)[3] to represent our data models. There are many other ways to represent data structure (for example, Entity Relationship Diagrams) that, for the purposes of this book, would also be suitable. We choose to use UML as it has a large suite of diagramming tools for developing software applications that encompasses not only the structure of data but also its behavior. In this section, we will look at how to interpret a class diagram and how to translate it into tables and constraints in a relational database.

A *class* is like a template for something we want to keep data about (events, people, places, etc.) For example, we might want to keep names and other details about the members of our golf club. Figure 1-5 shows the UML notation for a Member class. The name of the class is in the top panel, and the next panel shows the *attributes*. Class diagrams can also have another panel to show methods associated with the behavior of the class.

---

[3]If you want more information about UML, then refer to Grady Booch, James Rumbaugh, and Ivar Jacobsen, *The Unified Modeling Language User Guide* (Boston, MA: Addison Wesley, 2005). The current standards can be found at http://www.uml.org/.

Figure 1-5. *UML representation of a Member class*

In a relational database, each class is represented as a table, the attributes are the columns, and each instance (in this case an individual club member) will be a row in the table.

The data model can also depict the way the different classes depend on each other. Figure 1-6 shows two classes, Member and Team, and how they are related.

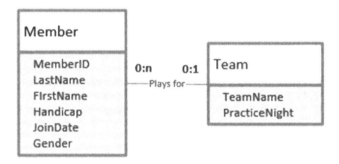

Figure 1-6. *A relationship between two classes*

The pair of numbers at each end of the *plays for* line in Figure 1-6 indicates how many members play for one particular team, and vice versa. The first number of each pair is the minimum number. This is often 0 or 1 and is therefore sometimes known as the *optionality* (that is, it indicates whether a member *must* have an associated team, or vice versa). The second number (known as the *cardinality*) is the greatest number of related objects. It is usually 1 or many (denoted by *n* or *), although other numbers are possible.

Relationships can be interpreted in both directions. The label on the relationship in Figure 1-6 implies that we are reading from left to right and we will need to think of the appropriate verb for interpreting the diagram in the other direction. "Team *has* members" will do. Reading Figure 1-6 from left to right, we see that one particular member doesn't have to *play for* a team and can *play for* at most one team (the numbers 0 and 1 at the end of the line nearest the Team class). Reading from right to left, we can say that one particular team doesn't need to *have* any members but can *have* many (the numbers 0 and n nearest the Member class). A relationship like the one in Figure 1-6 is called a 1–Many relationship (a member can belong to just one team, and a team can have many members).

You might think there should be exactly four members for a team (say, for an interclub team). Although this might be true when the team plays a round of golf, our database might record different numbers of members associated with the team as we add and remove players throughout the year. A data model usually uses 0, 1, and many to model the relationships between tables. Other constraints (such as the maximum number on a team) are more usually expressed with business rules or with UML use cases.[4]

We can represent a 1-Many relationship in our database by looking at the primary key at the 1 end of the relationship and adding a column of the same type to the table at the Many end. For the model in Figure 1-6 we would add a Team column to the Member table as shown in Figure 1-7.

MemberID ⊣	LastName ⊣	FirstName ⊣	Handicap ⊣	JoinDate ▾	Gender ▾	Team ▾
118	McKenzie	Melissa	30	28-May-05	F	
138	Stone	Michael	30	31-May-09	M	
153	Nolan	Brenda	11	12-Aug-06	F	TeamB
176	Branch	Helen		06-Dec-11	F	
178	Beck	Sarah		24-Jan-10	F	
228	Burton	Sandra	26	09-Jul-13	F	
235	Cooper	William	14	05-Mar-08	M	TeamB
239	Spence	Thomas	10	22-Jun-06	M	
258	Olson	Barbara	16	29-Jul-13	F	
286	Pollard	Robert	19	13-Aug-13	M	TeamB
290	Sexton	Thomas	26	28-Jul-08	M	
323	Wilcox	Daniel	3	18-May-09	M	TeamA

Figure 1-7. *Member table with a foreign key column Team*

The Team column is called a *foreign key*. Any non-empty value in this column in the Member table must be a value that already exists in the primary key column of the Team table. The concept of a foreign key provides us with a constraint on the Member table so that we cannot assign members to non-existent teams. This constraint is called *referential integrity*.

The SQL to create a table with a foreign key is shown here:

```
CREATE TABLE Member(
MemberID INT PRIMARY KEY,
LastName CHAR(20),
FirstName CHAR(20),
Phone CHAR(20),
Handicap INT,
JoinDate DATETIME,
Gender CHAR(1),
Team CHAR(20) FOREIGN KEY REFERENCES Team);
```

Because we need to compare the value in the foreign key column of the Member table with the primary key column of the Team table, these two columns must have the same domain or datatype.

Most database products have a graphical interface for setting up and displaying foreign key constraints. Figure 1-8 shows the interfaces for Microsoft© SQL Server and Microsoft© Access. These diagrams, which are essentially implementations of the data model, are invaluable for understanding the structure of the database so we know how to extract the information we require.

---

[4]Alistair Cockburn, *Writing Effective Use Cases* (Boston, MA: Addison Wesley, 2001).

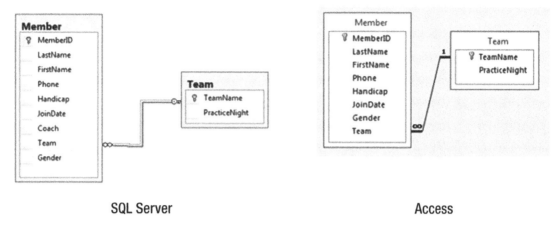

SQL Server   Access

Figure 1-8. *Diagrams for implementing 1–Many relationships using foreign keys*

The tables in Figures 1-4 and 1-7 have essentially the same design. For Figure 1-4 we arrived at the design by removing the PracticeNight column from the Member table and creating a new Team table (a normalization process). For Figure 1-7 we first considered a data model and added the Team column to the Member table as a way of representing the relationship between Member and Team. The outcome is the same whichever way you approach the issue.

At the risk of repeating myself, I do want to caution about the necessity of ensuring that the database is properly designed. The simple model in Figure 1-6 is almost certainly quite unsuitable even for the tiny amount of data it contains. A real club will probably want to keep track of how the membership of teams evolves over the years. This will involve including information about seasons or years along with the team membership information. Some members might play for more than one team during a year if they are called in as a substitute. That information may or may not be necessary to retain. Designing a useful database is a tricky job and outside the scope of this book.[5]

Retrieving Information from a Database

Now that we have a well-designed database consisting of interrelated normalized tables, we can start to look at how to extract information by way of queries. When I refer to extracting or retrieving information I don't mean that we are removing any data. Think of a query as providing a window onto a small part of the database. Many database systems will have a diagrammatic interface that can be useful for simple queries. Figure 1-9 shows the Microsoft© Access interface for retrieving the names of senior members from the Member table. The checkmarks denote which columns we want to retrieve, and the Criteria row enables us to specify conditions on the rows that are returned.

---

[5]For more information about database design, refer to my other Apress book, *Beginning Database Design: From Novice to Professional* (New York: Apress, 2012).

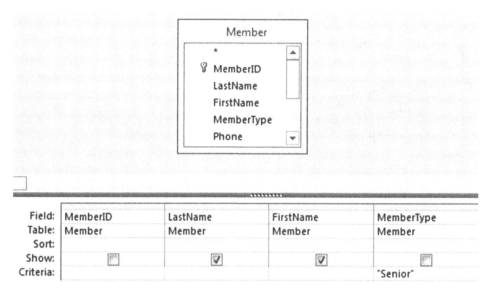

Figure 1-9. *Access interface for a simple query on the Member table*

The application will take the information from the graphical interface and construct an SQL query. Most applications will show you the SQL that is generated, and you can amend it or write it from scratch yourself. The SQL equivalent to the query depicted in Figure 1-9 is:

```
SELECT FirstName, LastName
FROM Member
WHERE MemberType = 'Senior';
```

This SQL query contains three clauses: SELECT specifies which columns to return, FROM specifies the table(s) where the information is kept, and WHERE specifies the conditions the returned rows must satisfy. We'll look at the structure of SQL statements in more detail later, but for now the intention of the query is pretty clear.

As we need to include more and more tables connected in a variety of ways, the diagrammatic interfaces rapidly become unwieldy, and often we need to write the SQL commands directly. Often, it is easier to think about a query in a more abstract way. With a clear abstract understanding of what is required, it then becomes more straightforward to turn the idea into an appropriate SQL statement. There are two different ways to approach queries on a relational database.

Process Approach

One way to approach a query is to think in terms of the operations we need to carry out on the tables. Let's think about how we might to get a list of names for members who practice on a Monday. We might imagine first retrieving just the rows from the Team table that have Monday in the PracticeNight column. We might then join those rows with the Member table (more about joins later) and then extract the names from the result. We will call this the *process approach*, as it is a series of steps carried out in a particular order. Figure 1-10 depicts the steps just described.

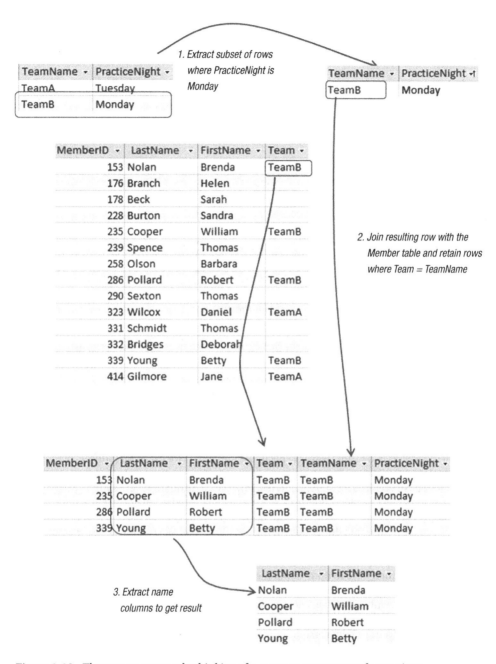

Figure 1-10. *The process approach: thinking of a query as a sequence of operations*

Outcome Approach

An alternative way to think about the query in the previous section is to examine all the rows in the Member table and just return those that satisfy the criteria that the member is on a team that has Monday as a practice night. Figure 1-11 depicts this train of thought. The row m that we are considering in the Member table satisfies the condition about the team's practice night, so we should retrieve the names from that row.

11

MemberID ▾	LastName ▾	FirstName ▾	Team ▾
m ☞ 153 Nolan		Brenda	(TeamB)
176 Branch		Helen	
178 Beck		Sarah	
228 Burton		Sandra	
235 Cooper		William	TeamB
239 Spence		Thomas	
258 Olson		Barbara	
286 Pollard		Robert	TeamB
290 Sexton		Thomas	
323 Wilcox		Daniel	TeamA
331 Schmidt		Thomas	
332 Bridges		Deborah	
339 Young		Betty	TeamB
414 Gilmore		Jane	TeamA

TeamName ▾	PracticeNight ▾
TeamA	Tuesday
t ☞ (TeamB)	Monday

Figure 1-11. *Considering if the row m satisfies the criteria for the query.*

We will call this type of thinking about a query the *outcome approach* because we describe what we want rather than how to get it.

Why We Consider Two Approaches

Relational database theory has its origins in set theory. If we think of our tables as sets of rows, then a query is a question that requires us to manipulate those sets to retrieve a subset containing the information we require. The relational theory has two formal ways of specifying the criteria for extracting subsets of rows: relational algebra and relational calculus.

We do not need these abstract ideas for simple queries. However, if all queries were simple, you would not be reading this book. In the first instance, queries are expressed in everyday language that is often ambiguous. Try this simple expression: "Find me all students who are younger than 20 or live at home and get an allowance." This can mean different things depending on where you insert commas. For example a comma after "20" leads to the interpretation that everyone under 20 is included, while a comma after "home" suggests that they must also get an allowance. Even after we have sorted out what the natural-language expression means, we then have to think about the query in terms of the actual tables in the database. This means having to be quite specific in how we express the query. Both relational algebra and relational calculus give us a powerful way of being accurate and specific.

Why not skip all this abstract stuff and go right ahead and learn SQL? Well, the SQL language consists of elements of both calculus and algebra. Older versions of SQL were purely based on relational calculus in that you described *what* you wanted to retrieve rather than *how*. Modern implementations of SQL allow you to explicitly specify algebraic operations such as joins, unions, and intersections on the tables as well.

There are often several equivalent ways of expressing an SQL statement. Some ways are very much based on calculus, some are based on algebra, and some are a bit of both. During my time as a university lecturer I often asked the class whether they found the calculus or algebra expressions more intuitive for a particular query. The class was usually equally divided. Personally, I find that some queries just feel obvious in terms of relational algebra, whereas others feel much more simple when expressed in relational calculus. Once I have the idea pinned down with one or other, the translation into SQL (or some other query language) is usually straightforward.

We can make use of the ideas of relational algebra and relational calculus without delving into the mathematics. In the body of the book I refer to the *process approach* (algebra) and the *outcome approach* (calculus). The more tools you have at your disposal, the more likely it is that you will be able to express complex queries accurately. In Appendix 2 there is an introduction to the formal notation for relational algebra and relational calculus for those of you who would like to add that to your armory.

Summary

This chapter has presented an overview of relational databases. We have seen that a relational database consists of a set of tables that represent the different aspects of our data (for example, a table for members and a table for teams). The attributes needed to describe the members or teams become the columns of the tables, and each column has a set of allowed values (a domain). Each table should have a primary key, which is an attribute or set of attributes guaranteed to have a different value for every row.

It is possible to set up constraints between tables with foreign keys. A foreign key is a value for a column(s) in one table that has to already exist as a value in the primary key column(s) of another table. For example, the value of Team in the Member table must be one of the values in the primary key field of the Team table.

It is often helpful to think about queries in an abstract way, and there are two ways to do this. The process approach requires us to think about the operations that can be applied to tables in a database. It is a way of describing *how* we need to manipulate the tables to extract the information we require. The outcome approach requires us to think about *what* criteria our required information must satisfy. Different people will find that one or the other of these approaches feels more natural for different queries. SQL is a language for specifying queries on a database. There are usually many equivalent ways to specify a query in SQL. Some reflect the process approach and some reflect the outcome approach – and some are a bit of both.

CHAPTER 2

▨ ▨ ▨

Simple Queries on One Table

If a database has been designed correctly, the data will be located in several different tables. For example, our golf database has separate tables for information about members, teams, and tournaments, as well as tables that connect these values; for example, which members play on which teams, enter which tournaments, and so on. To make the best use of our data, we will need to inspect values from different tables to retrieve the information we require.

In this chapter, we will look at retrieving information from a single table. The table may be one of the permanent tables in the database, or it may be a virtual table that has been temporarily put together as part of a more complicated query.

I've been talking in a rather imprecise manner about "retrieving" rows and "returning" information. What happens to the rows that result from a query? In reality, we are not removing data from tables and putting it somewhere. A query is like a window onto the database through which we can see just the information we require. If the data in the underlying database changes, then the results of our query will change too. It doesn't hurt to think about the information that results from a query as being "retrieved" into a "virtual" table as long as you realize it is just temporary.

Subsets of Rows and Columns

Selecting subsets of rows and/or columns is one of the most common operations we will carry out in a query. In the following sections, we will look at selecting rows and columns from one of the original tables in the database.[1] The same ideas apply to retrieving information from virtual tables that result from other manipulations of the data.

To determine which rows to retrieve from a table, it is necessary to specify a *condition*, which is a statement that is either true or false. We apply the condition to each row in the table independently, retaining those rows for which the condition is true and discarding the others. Say we want to find all the seniors in the golf club. We want just that subset of rows from the Member table where the value in the MemberType field is "Senior," as shown in Figure 2-1.

---

[1]In the formal terms of relational algebra, retrieving a subset of rows (tuples) from a table (relation) is known as the *select* operation and retrieving a subset of attributes (columns) is known as the *project* operation. See Appendix 2 for more information.

© Clare Churcher 2016

C. Churcher, *Beginning SQL Queries*, DOI 10.1007/978-1-4842-1955-3_2

MemberID	LastName	FirstName	MemberType	Phone	Handicap	JoinDate	Coach	Team	Gender
118	McKenzie	Melissa	Junior	963270	30	28/05/2005	153		F
138	Stone	Michael	Senior	983223	30	31/05/2009			M
153	Nolan	Brenda	Senior	442649	11	12/08/2006		TeamB	F
176	Branch	Helen	Social	589419		6/12/2011			F
178	Beck	Sarah	Social	226596		24/01/2010			F
228	Burton	Sandra	Junior	244493	26	9/07/2013	153		F
235	Cooper	William	Senior	722954	14	5/03/2008	153	TeamB	M
239	Spence	Thomas	Senior	697720	10	22/06/2006			M
258	Olson	Barbara	Senior	370186	16	29/07/2013			F
286	Pollard	Robert	Junior	617681	19	13/08/2013	235	TeamB	M
290	Sexton	Thomas	Senior	268936	26	28/07/2008	235		M
323	Wilcox	Daniel	Senior	665393	3	18/05/2009		TeamA	M
331	Schmidt	Thomas	Senior	867492	25	7/04/2009	153		M
332	Bridges	Deborah	Senior	279087	12	23/03/2007	235		F
339	Young	Betty	Senior	507813	21	17/04/2009		TeamB	F
414	Gilmore	Jane	Junior	459558	5	30/05/2007	153	TeamA	F
415	Taylor	William	Senior	137353	7	27/11/2007	235	TeamA	M
461	Reed	Robert	Senior	994664	3	5/08/2005	235	TeamA	M
469	Willis	Carolyn	Junior	688378	29	14/01/2011			F
487	Kent	Susan	Social	707217		7/10/2010			F

Figure 2-1. *Retrieving the subset of rows for Senior members.*

The SQL for the query to retrieve Senior members is as follows:

```
SELECT *
FROM Member
WHERE MemberType = 'Senior'
```

This query has three parts, or *clauses*: The SELECT clause says what columns to retrieve. In this case, * means retrieve all the columns. The FROM clause says which table(s) the query involves, and the WHERE clause describes the condition for deciding whether a particular row should be included in the result. The condition says to check the value in the field MemberType. In SQL, when we specify an actual value for a character or text field, we need to enclose the value in single quotes, as in 'Senior'.

Now let's look at how we can specify that we want to see only some of the columns in our result. I will generally refer to *selecting* a subset of rows and *projecting* a subset of columns. Often the projection of a subset of columns is the last step in a series of operations. We can think of gathering all the data we require and then at the end asking for just the attributes or columns we need. We will see in Chapter 7 that we sometimes also need to project similar columns from original or virtual tables before applying some of the set operations, such as union and intersection.

If we want a phone list of all the members we don't need extra information such as handicaps or join dates. Figure 2-2 show a subset of the name and phone number columns from the Member table.

MemberID	LastName	FirstName	MemberType	Phone	Handicap	JoinDate	Coach	Team	Gender
118	McKenzie	Melissa	Junior	963270	30	28/05/2005	153		F
138	Stone	Michael	Senior	983223	30	31/05/2009			M
153	Nolan	Brenda	Senior	442649	11	12/08/2006		TeamB	F
176	Branch	Helen	Social	589419		6/12/2011			F
178	Beck	Sarah	Social	226596		24/01/2010			F
228	Burton	Sandra	Junior	244493	26	9/07/2013	153		F
235	Cooper	William	Senior	722954	14	5/03/2008	153	TeamB	M
239	Spence	Thomas	Senior	697720	10	22/06/2006			M
258	Olson	Barbara	Senior	370186	16	29/07/2013			F
286	Pollard	Robert	Junior	617681	19	13/08/2013	235	TeamB	M
290	Sexton	Thomas	Senior	268936	26	28/07/2008	235		M
323	Wilcox	Daniel	Senior	665393	3	18/05/2009		TeamA	M
331	Schmidt	Thomas	Senior	867492	25	7/04/2009	153		M
332	Bridges	Deborah	Senior	279087	12	23/03/2007	235		F
339	Young	Betty	Senior	507813	21	17/04/2009		TeamB	F
414	Gilmore	Jane	Junior	459558	5	30/05/2007	153	TeamA	F
415	Taylor	William	Senior	137353	7	27/11/2007	235	TeamA	M
461	Reed	Robert	Senior	994664	3	5/08/2005	235	TeamA	M
469	Willis	Carolyn	Junior	688378	29	14/01/2011			F
487	Kent	Susan	Social	707217		7/10/2010			F

Figure 2-2. *Projecting a subset of columns to provide a phone list*

The SQL to retrieve the name and phone columns from the Member table is:

```
SELECT LastName, FirstName, Phone
FROM Member
```

Because we want to see these column values for *every* row, this query doesn't have a WHERE clause.

It is a simple matter to combine the retrieval of subsets of rows and columns. We might do this if we wanted a phone list for just the senior members, as in Figure 2-3.

MemberID	LastName	FirstName	MemberType	Phone	Handicap	JoinDate	Coach	Team	Gender
118	McKenzie	Melissa	Junior	963270	30	28/05/2005	153		F
138	Stone	Michael	Senior	983223	30	31/05/2009			M
153	Nolan	Brenda	Senior	442649	11	12/08/2006		TeamB	F
176	Branch	Helen	Social	589419		6/12/2011			F
178	Beck	Sarah	Social	226596		24/01/2010			F
228	Burton	Sandra	Junior	244493	26	9/07/2013	153		F
235	Cooper	William	Senior	722954	14	5/03/2008	153	TeamB	M
239	Spence	Thomas	Senior	697720	10	22/06/2006			M
258	Olson	Barbara	Senior	370186	16	29/07/2013			F
286	Pollard	Robert	Junior	617681	19	13/08/2013	235	TeamB	M
290	Sexton	Thomas	Senior	268936	26	28/07/2008	235		M
323	Wilcox	Daniel	Senior	665393	3	18/05/2009		TeamA	M
331	Schmidt	Thomas	Senior	867492	25	7/04/2009	153		M
332	Bridges	Deborah	Senior	279087	12	23/03/2007	235		F
339	Young	Betty	Senior	507813	21	17/04/2009		TeamB	F
414	Gilmore	Jane	Junior	459558	5	30/05/2007	153	TeamA	F
415	Taylor	William	Senior	137353	7	27/11/2007	235	TeamA	M
461	Reed	Robert	Senior	994664	3	5/08/2005	235	TeamA	M
469	Willis	Carolyn	Junior	688378	29	14/01/2011			F
487	Kent	Susan	Social	707217		7/10/2010			F

Figure 2-3. *Retrieving a subset of rows and columns to produce a phone list of Senior members*

The SQL for the query depicted in Figure 2-3 is:

```
SELECT LastName, FirstName, Phone
FROM Member
WHERE MemberType = 'Senior'
```

Using Aliases

As our queries get more complicated they will incorporate a number of different tables. Some of the tables may have the same column names, and we might need to distinguish them from each other. In SQL we can preface each of the attributes in our query with the name of the table that it comes from, as shown here:

```
SELECT Member.LastName, Member.FirstName, Member.Phone
FROM Member
WHERE Member.MemberType = 'Senior'
```

Because typing the whole table name can become tiresome, and also because in some queries we might need to compare data from more than one row of a table, SQL has the notion of an *alias*. Have a look at the following query:

```
SELECT m.LastName, m.FirstName, m.Phone
FROM Member m
WHERE m.MemberType = 'Senior'
```

In the FROM clause, we have declared an alias or alternative name for the Member table, in this case m. We can give our alias any name or letter we like; shorter is better. Then, in the rest of the query we can use the alias whenever we want to specify an attribute from that table. It is a good idea to get into the habit of using a table alias for each table contributing to the query.

Saving Queries

It is possible to keep the result of a query in a new permanent table (sometimes called a *snapshot*), but we usually don't want to do that because it will become out of date if the underlying data changes. What we usually want to do is save the query instructions so that we can ask the same question another day. Consider our phone list query. Every so often after the membership of the club has been updated, we will produce a new phone list. Rather than having to construct the query each time, we can save the instructions in what is known as a *view*. The code below shows how to create a view that we can use to provide up-to-date phone lists. We have to give the view a name, which can be anything we want (PhoneList seems sensible), and then we supply the SQL statement for retrieving the appropriate data:

```
CREATE VIEW PhoneList AS
SELECT m.LastName, m.FirstName, m.Phone
FROM Member m
```

You can think of PhoneList as the instructions to create a "virtual" table that we can use in other queries in the same way that we use real tables. We just need to remember that the virtual table is created on the fly by running the query on the permanent Member table and it is then gone. To get our phone list now, we can simply use the PhoneList view:

```
SELECT * FROM PhoneList
```

Specifying Conditions for Selecting Rows

In the queries we looked at in the previous sections, we used very simple conditions or criteria for determining whether to include a row in the result of a query. In the following section, we will look more closely at the different ways you can specify more complicated conditions.

Comparison Operators

A *condition* is a statement or expression that is either true or false, such as MemberType = 'Senior'. These types of expressions are called *Boolean expressions* after the 19th-century English mathematician, George Boole, who investigated their properties. The conditions we use to select rows from a table usually involve comparing the values of an attribute to some constant value or another attribute. For example, we can ask whether the value of an attribute is the same, different, or greater than some value. Table 2-1 shows some comparison operators we can use in our queries.

Table 2-1. *Comparison Operators*

Operator	Meaning	Examples of True Statement
=	Equals	5=5, 'Junior' = 'Junior'
<	Less than	4<5, 'Ann' < 'Zebedee'
<=	Less than or equal to	4<=5, 5<=5
>	Greater than	5>4, 'Zebedee' > 'Ann'
>=	Greater than or equal to	5>=4, 5>=5
<>	Not equal	5<>4, 'Junior' <> 'Senior'

Just a quick note of caution: in Table 2-1, some of our examples compare numbers, and some compare characters. Recall from Chapter 1 that when we create a table, we specify the type of each field; for example, MemberID was declared to be an INT (integer or whole number), and LastName a CHAR(20) (a 20-character field). With fields like integer, comparisons are numerical. With text or character fields, comparisons are alphabetical, and with date and time fields, comparisons are chronological (earlier dates come first).

When we compare character attributes, the comparison is based on the ASCII[2] or Unicode value of the characters. As we might expect "A" (ASCII value 65) comes before "Z" (ASCII 90), so "A" < "Z". With a string of characters, if the first letter is the same then the order is decided by the second, and so on. So "ANNABEL" < "ANNE". However, the lowercase characters have higher ASCII codes than the uppercase ones. This means that "a" (ASCII 97) > "Z" (ASCII 90). If you order a list of names alphabetically then, by default, a name starting with a lowercase letter will appear after those starting with uppercase letters. For example "van Dyke" will appear after "Zebedee."

If we put numbers in a character field, they will also sort alphabetically. This means you will have comparisons such as "400" < "5", because the first character, "4" (ASCII 34), in the left-hand text is less than the first character, "5" (ASCII 35), on the right-hand side. So, make sure if a column is going to contain numbers that you want to compare and order numerically, that it is declared as a numeric type, or you will get some rather surprising results from your queries. Similarly, dates need to be in a column declared with one of the date types or the comparisons and ordering may not be what you expect.

With comparison operators, we can create many different queries. Table 2-2 shows some examples of Boolean expressions that we can use as conditions in the WHERE clause of an SQL statement for selecting rows from the Member table.

Table 2-2. *Examples of Boolean Expressions on the Member Table*

Expression	Retrieved Rows
MemberType = 'Junior'	All junior members
Handicap <= 12	All members with a handicap of 12 or less
JoinDate >= '01/01/2008'	Everyone who has joined after the beginning of 2008
Gender = 'F'	All the women

---

[2]http://www.asciitable.com/

Some implementations of SQL are case sensitive when comparing text, and others are not. Being case sensitive means that uppercase letters are treated as being different from their lowercase counterpart; in other words, "Junior" is different from "junior," which is different from "JUNIOR." I usually check out any new database system I use to see what it does. If you do not care about the case of the attribute you are considering (that is, you are happy to retrieve rows where MemberType is "Junior" or "jUnIoR" or whatever), you can make use of the SQL function UPPER. This will turn the value of each text attribute into uppercase before you do the comparison. You can then compare that with the uppercase literal value, as shown here:

```
SELECT *
FROM Member m
WHERE UPPER(m.MemberType) = 'JUNIOR'
```

Logical Operators

We can combine Boolean expressions to create more interesting conditions. For example, we can specify that two expressions must both be true before we retrieve a particular row.

Let's assume we want to find all the junior girls. This requires two conditions to be true: they must be female, and they must be juniors. We can easily express each of these conditions independently. After that, we can use the logical operator AND to require that *both* conditions be true:

```
SELECT *
FROM Member m
WHERE m.MemberType = 'Junior' AND m.Gender = 'F'
```

We will look at three logical operators: AND, OR, and NOT. We have already seen how AND works. If we use OR between two expressions, then only one of the expressions need be true (but if they are both true, that is OK as well). NOT is used before an expression. For example, for our Member table, we might ask for rows obeying the condition NOT (MemberType = 'Social'). This means check each row, and if the value of MemberType is "Social," then we do *not* want that row. Table 2-3 gives some more examples of using logical operators in conditions.

Table 2-3. *Examples of Logical Operators*

Expression	Description of Data
MemberType = 'Senior' AND Handicap < 12	Seniors with a handicap under 12
MemberType = 'Senior' OR Handicap < 12	All the senior members as well as anyone else with a good handicap (those less than 12)
NOT(MemberType = 'Social')	All the members except the social ones (for the current data, that would be just the seniors and juniors)

Figure 2-4 shows a diagrammatic representation of the queries in Table 2-3. Each circle represents a set of rows (that is, those for social members or those for members with handicaps under 12). The shaded area represents the result of the operation.

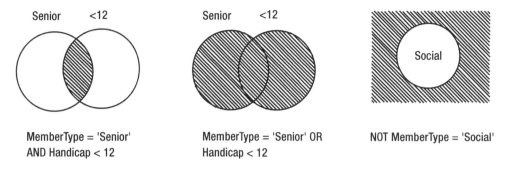

Figure 2-4. Diagrammatic representation of the logical operators.

The truth tables in Figure 2-5 can be helpful in understanding how the logical operators work. You read them like this: in Figures 2-5a and 2-5b, we have two expressions, one along the top and one down the left. Each expression can have one of two values: True (T) or False (F). If we combine them with the Boolean expression AND, then Figure 2-5a shows that the overall statement is true only if both the contributing statements are true (the square in the top left). If we combine them with an OR statement, then the overall statement is false only if both contributing statements are false (bottom right of Figure 2-5b). The table in Figure 2-5c says that if our original statement is true and we put NOT in front, then the result is false (left column), and vice versa.

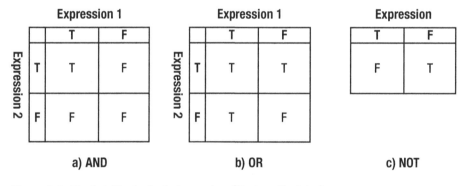

Figure 2-5. Truth tables for logical operators (T = true, F = false)

Sometimes it can be a bit tricky turning natural-language descriptions into Boolean expressions. If you were asked for a list that included *all the women and all the juniors* (don't ask why!), you might translate this literally and write the condition MemberType = 'Junior' AND Gender = 'F'. However, the AND means *both* conditions must be true, so this would give us junior women. What our natural-language statement really means is "I want the row for any member if they are either a woman *or* a junior (or both)." Be careful.

Dealing with Nulls

The example data in the Member table shown earlier in Figure 2-1 is all accurate and complete. Every row has a value for each attribute, except for Handicap, which doesn't apply to some members. Real data is usually not so clean and tidy. Let's consider some different data, as in Figure 2-6.

LastName ▾	FirstName ▾	MemberType ▾	Handicap ▾	Gender ▾	JoinDate ▾
McKenzie	Melissa	Junior	30	F	28-May-05
Stone	Michael	Senior	30	M	
Nolan	Brenda	Senior	11	F	12-Aug-06
Branch	Helen	Social		F	06-Dec-11
Beck	Sarah			F	24-Jan-10
Burton	Sandra	Junior	26	F	09-Jul-13
Cooper	William	Senior	14	M	05-Mar-08
Spence	Kim	Senior	10		22-Jun-06
Olson	Barbara	Senior	16	F	29-Jul-13
Pollard	Robert	Junior	19	M	13-Aug-13
Sexton	Thomas	Senior	26	M	28-Jul-08
Wilcox	Daniel	Senior	3	M	18-May-09

Figure 2-6. *Table with missing data*

When there is no value in a cell in a table, it is said to be *null*. Nulls in a database can cause a few headaches. Consider carrying out the following two queries: one to produce a list of male members and the other a list of females. Given that golfers need to identity as either male or female for competition purposes, we might assume that all the members of the club would appear on one list or the other. However, for the data in Figure 2-6, we would leave out Kim Spence. You could argue that the data shouldn't be like that, but we are talking about real people and real clubs with less than accurate and complete data. Maybe Kim forgot (or refused) to fill in the gender part of the application form. We can protect against this by insisting that nulls are not allowed in a particular field when we create a table. The following SQL statement shows how we could make Gender a field that always requires a value:

```
CREATE TABLE Member (
MemberID INT PRIMARY KEY,
.....
Gender CHAR(1) NOT NULL,

....)
```

It is worth bearing in mind that making fields NOT NULL can create more headaches than it cures. If Kim Spence did not complete all the boxes on his/her membership application but had organized payment for the subscription, then we want to record him/her as a member and worry about the full details later. If we make Gender a required field, then we can't enter a record for him/her in the table–or we have to guess what his/her gender is. Neither of these options is a good strategy, so it is best to be sparing when making fields required. Remember that our primary key fields (by definition) always need a value.

Not all values of null mean there is a problem with the data. In our Member table, a field might be null because it does not apply to a particular member. Helen and Sarah's handicap may be genuinely null because they do not have handicaps. However, it is fair to assume that every member should have a value for MemberType and JoinDate, so the nulls in these columns are because we do not know the value. In the real world, expect that your tables will have missing data.

Finding Nulls

Given that in our tables we may have nulls that might cause us problems, it is useful to be able to find them. After we have entered a batch of new members into the database, we can check for problems. We might want to get a list of all the members who don't have a value for Gender, say. To do this we can use the SQL phrase IS NULL:

```
SELECT *
FROM Member m
WHERE m.Gender IS NULL
```

Alternatively, we might want to retrieve only those members who *do* have a value in a cell. If we want the names and handicaps of only those members who have a value for Handicap, we could use the NOT operator to create the following query:

```
SELECT *
FROM Member m
WHERE NOT (m.Handicap IS NULL)
```

Comparisons Involving Null Values

Given that we are going to have unexpected nulls in our tables, it is important to know how to deal with them. What rows will match the two conditions shown here?

```
Gender = 'F'
NOT (Gender = 'F')
```

You might think that if we carry out two queries, one to get all the rows that match a condition and another for all the rows that don't match, then we will get the whole table. But, in fact, we don't. Kim will not be included with the first condition, because clearly the value of Gender does not equal 'F'. But when we ask whether the value is NOT 'F' we can't say, because we don't know what the value is. It might be 'F' if it had a value. In SQL when we compare null values with something, we don't get either True or False because we simply don't know. This probably makes more sense if we think about handicaps. If we ask for everyone with Handicap > 12, and also for those members who satisfy either NOT (Handicap > 12) or Handicap <=12, then Sarah's row will never be retrieved. The question doesn't apply to her – she doesn't have a handicap.

Once we take nulls into consideration, our expressions for conditions might actually have one of three values: True, False, or "Don't know." That is pretty much how the world works, if you think about it. Only rows that are True for a condition are retrieved in a query. If the condition is False or if we don't know, then the row is not retrieved.

If we include "Don't know" in the truth tables they will look like those in Figure 2-7. For an AND operation, if one expression is False, then it doesn't matter about the others – the result will be False. For an OR operation, if one expression is True, then it doesn't matter about the others, so the result will be True.

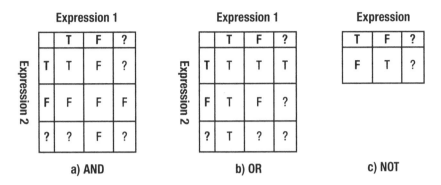

Figure 2-7. *Truth tables with three-valued logic (T = True, F = False, ? = Don't know)*

Managing Duplicates

If our tables have been designed well, they will have a primary key. This ensures that every row is unique. However, as soon as we retrieve a subset of data from the tables the result may not have unique rows.[3] Let's look at an example.

Consider retrieving just the FirstName column from the Member table. Figure 2-8 shows two possible results.

---

[3]Formally, in terms of relational algebra, the result of every operation will generate another relation or set of unique rows. See Appendix 2 for more information.

FirstName ▾
Melissa
Michael
Brenda
Helen
Sarah
Sandra
William
Thomas
Barbara
Robert
Thomas
Daniel
Thomas
Deborah
Betty
Jane
William
Robert
Carolyn
Susan

FirstName ▾
Barbara
Betty
Brenda
Carolyn
Daniel
Deborah
Helen
Jane
Melissa
Michael
Robert
Sandra
Sarah
Susan
Thomas
William

a) With duplicates   b) Without duplicates

Figure 2-8. *Projecting the FirstName column from the Member table*

It is useful to think about why we might carry out a query retrieving just names. Perhaps the query is to help prepare a set of nametags for a club party. If that is the case, then two Thomases and a William are going to feel left out if we use the unique output.

You might think, what's all the fuss? Of course we want to keep all the rows. However, consider retrieving just the column with the membership types. Figure 2-9 shows the outputs with duplicates included and removed.

MemberType ▾
Junior
Senior
Senior
Social
Social
Junior
Senior
Senior
Senior
Junior
Senior
Senior
Senior
Senior
Senior
Junior
Senior
Senior
Junior
Social

MemberType ▾
Junior
Senior
Social

a) With duplicates   b) Without duplicates

Figure 2-9. *Projecting the MemberType column from the Member table*

It's pretty difficult to think of a situation where you want the duplicated rows in Figure 2-9a. The two operations we have considered sound similar in natural language. "Give me a list of first names" and "Give me a list of membership types" sound like the same sort of question, but they mean quite different things. The first means "Give me a name for each member," and the other means "Give me a list of unique membership types."

What does SQL do? If we say SELECT MemberType FROM Member, we will get the output in Figure 2-9a with all the duplicates included. If we do not want the duplicates, then we can use the keyword DISTINCT:

```
SELECT DISTINCT m.MemberType
FROM Member m
```

Whether or not you keep the duplicates depends very much on the information you require, so you need to give it careful thought. If you were expecting the set of rows in Figure 2-9b and got Figure 2-9a, you would most likely notice. With the two sets of rows in Figure 2-8, it is much more difficult to spot that you have perhaps made a mistake. Get into the habit of thinking about duplicates for all your queries.

Ordering Output

Every now and then I refer to a "set of rows" rather than a table or a virtual table. The word *set* has two implications. One is that there are no duplicates (and we have discussed that a lot!). The other implication is that there is no particular order to the rows in our set. In theory, we don't have a first row or a last row or a next row. If we run a query to retrieve all the rows, or just some of the rows, from a table, then we have no guarantee in what order they will be returned. However, sometimes we might like to display the results in a particular order. We can do this with the key phrase ORDER BY. The following shows how to retrieve member information ordered alphabetically by LastName:

```
SELECT *
FROM Member m
ORDER BY m.LastName
```

We can order by two or more values. For example, if we want to order Senior members with the same LastName by the value of their FirstName, we can include those two attributes (in that order) in the ORDER BY clause:

```
SELECT *
FROM Member m
WHERE m.MemberType = 'Senior'
ORDER BY m.LastName, m.FirstName
```

The type of a field determines how the values will be ordered. By default, text fields will be ordered alphabetically, number fields will be ordered numerically (smallest first), and date and time fields chronologically (earlier dates and times first). We can also specify that the order be reversed with the keyword DESC (for descending). There is an equivalent keyword ASC (for ascending), which is the default if neither is specified. The following will return member names and handicaps ordered in descending order; i.e., with the highest value of handicap first:

```
SELECT m.Lastname, m.FirstName, m.Handicap
FROM Member m
ORDER BY m.Handicap DESC
```

The way nulls are ordered in any output depends on the application; you will need to check. For example, in SQL Server and Microsoft Access, nulls will appear at the top of an ascending list and the bottom of a descending list. Oracle provides keywords such as NULLS FIRST and NULLS LAST so you can choose where the null values go. A little trick to get your nulls at the bottom of an ascending list in SQL Server is to use a case statement:

```
SELECT m.LastName, m.FirstName, m.Handicap
FROM Member m
ORDER BY (CASE
            WHEN m.Handicap IS NULL THEN 1
            ELSE 0
        END), m.Handicap
```

The preceding query has two attributes in the ORDER BY clause. It orders firstly by the case statement in the parentheses. You can think of the case statement as creating a virtual column giving the value 0 to those rows with a handicap and 1 to those which have no handicap value. When we order by this first attribute in the ORDER BY clause, the rows with a value for a handicap will be before the nulls. Within these groups the rows will then be ordered by the value of the handicap in ascending order.

Performing Simple Counts

As well as retrieving a subset of rows and columns from a table, we can also use SQL queries to provide some statistics. There are SQL functions that allow us to count records, total or average values, find maximum and minimum values, and so on. In this section, we will look at some simple queries for counting records. We will return to this topic in Chapter 8.

We can use the COUNT function to return the number of records in the Member table. In the following query, * means count each record:

```
SELECT COUNT(*) FROM Member
```

We can also count a subset of rows by adding a WHERE clause to specify those rows we want to include. For example, we can use the following query to count the number of senior members:

```
SELECT COUNT(*) FROM Member m
WHERE m.MemberType = 'Senior'
```

Because we have just been talking about nulls and duplicate values, it is worth briefly mentioning here how these will affect our counts. Rather than use * as a parameter to the COUNT function so that it counts all the rows, we can put an attribute such as Handicap in the parentheses. If we do this only those rows with a value in the Handicap field will be included in the count.

```
SELECT COUNT(Handicap) FROM Member
```

We can also specify that we want to count the number of unique values for an attribute. If we want to know how many different values of MemberType appear in the Member table then we can use the following query:

```
SELECT COUNT(DISTINCT MemberType) FROM Member
```

It is worth reiterating that different database software will support different parts of the SQL standard syntax. For example, Microsoft Access currently does not support COUNT(DISTINCT MemberType), seen in the previous query. There is usually a way to work around these differences to find an equivalent query, and we will look at how to rephrase the preceding query and other issues related to aggregates and summaries in Chapter 8.

Avoiding Common Mistakes

Retrieving a subset of rows and columns from a single table is the most simple of SQL queries. However, you have seen that you still need to be careful. It is important to remember that there will be null values in your tables and to think carefully about how your selection conditions will treat them. You also need to remember that if you do not retain the primary key fields from your tables, there is the potential to have duplicate rows, and you must deal with them appropriately.

There are a couple of other mistakes that are commonly made when selecting a subset of rows. They don't become apparent with a table like Member, so I'll introduce some more of the tables in the golf club database. Figure 2-10 shows part of the Member table and two other tables: Entry and Tournament. The first row in the Entry table records that person 118 (Melissa McKenzie) entered tournament 24 (Leeston) in 2014.

MemberID	LastName	FirstName
118	McKenzie	Melissa
138	Stone	Michael
153	Nolan	Brenda
176	Branch	Helen
178	Beck	Sarah
228	Burton	Sandra
235	Cooper	William
239	Spence	Thomas
258	Olson	Barbara
286	Pollard	Robert
290	Sexton	Thomas
323	Wilcox	Daniel
331	Schmidt	Thomas
332	Bridges	Deborah
339	Young	Betty
414	Gilmore	Jane
415	Taylor	William
461	Reed	Robert
469	Willis	Carolyn
487	Kent	Susan

a) Member (Some columns)

MemberID	TourID	Year
118	24	2014
228	24	2015
228	25	2015
228	36	2015
235	38	2013
235	38	2015
235	40	2014
235	40	2015
239	25	2015
239	40	2013
258	24	2014
258	38	2014
286	24	2013
286	24	2014
286	24	2015
415	24	2015
415	25	2013
415	36	2014
415	36	2015
415	38	2013
415	38	2015
415	40	2013
415	40	2014
415	40	2015

b) Entry

TourID	TourName
24	Leeston
25	Kaiapoi
36	WestCoast
38	Canterbury
40	Otago

c) Tournament

Figure 2-10. *Introducing the Tournament and Entry tables*

We can use some of the SQL operations we have already seen on the Entry table to answer questions such as which tournaments (just the TourID number) person 258 has entered, who (just the MemberID number) has ever entered tournament 24, or who entered tournament 36 in 2015. The following is the SQL for the last query:

```
SELECT e.MemberID
FROM Entry e
WHERE e.TourID = 36 AND e.Year = 2015
```

Incorrectly Using a WHERE Clause to Answer Questions with the Word "both"

In the previous section we used the logical operator AND to find rows in the Entry table where both TourID = 36 and Year = 2015 were true.

Say we wanted to find the members who have entered *both* tournaments 36 and 38. There is a temptation to again use the AND operator and write the query as follows:

```
SELECT e.MemberID
FROM Entry e
WHERE e.TourID = 36 AND e.TourID= 38
```

Can you work out what this query will return? This is where it is helpful to think in terms of the row variable e investigating each row in table Entry as in Figure 2-11.

MemberID ▾	TourID ▾	Year ▾
286	24	2014
286	24	2015
415	24	2015
415	25	2013
415	36	2014
415	36	2015
415	38	2013
415	38	2015
415	40	2013
415	40	2014
415	40	2015

(Note: the row "415 | 36 | 2015" is marked with the row variable e and a pointing-finger icon.)

Figure 2-11. *The row variable e investigates each row independently.*

Imagine our finger is pointing at the row shown in the diagram. Does this row (415, 36, 2015) satisfy the condition e.TourID = 36 AND e.TourID= 38? It satisfies the first part, but the AND operator requires the row to satisfy both conditions. No single row in our table will have *both* 36 and 38 in the tournament column because each row is for just one entry. The SQL query we suggested will never find any rows; it will always return an empty table. If we change the Boolean operator to OR, we will get the row indicated in Figure 2-10 returned; however, we will also then get anyone who has entered either 36 or 38 but not necessarily both.

This particular query cannot be solved with a simple WHERE clause. By definition, the condition in the WHERE applies to *each row independently*. To answer the question about who has entered *both* competitions, we need to look at more than one row of the Entry table at the same time (that is, two fingers). If we have two fingers, one pointing at the row shown in Figure 2-10 and another pointing at the following row, then we can deduce that 415 has entered both tournaments. We'll look at how to do this in Chapter 5.

Incorrectly Using a WHERE Clause to Answer Questions with the Word "not"

Now let's consider another common error. It is easy to find the people who have entered tournament 38 with the condition `e.TourID = 38`. It is tempting to try to retrieve the people who have *not* entered tournament 38 by changing the condition slightly. Can you figure out what rows the following SQL query will retrieve?

```
SELECT e.MemberID
FROM Entry e
WHERE e.TourID <> 38
```

What about the row that the finger is pointing to in Figure 2-11? Does this satisfy `e.TourID <> 38`? It certainly does. But this doesn't mean 415 hasn't entered tournament 38 (the following row says he did). The query, in fact, returns all the people who have entered some tournament that isn't tournament 38 (which is unlikely to be a question you'll ever want to ask!).

This is another type of question that can't be answered with a simple `WHERE` clause that looks at independent rows in a table. In fact, we can't even answer this question with a query that involves only the `Entry` table. Member 138, Michael Stone, has not entered tournament 38, but he doesn't even get a mention in the `Entry` table because he has never entered any tournaments at all. We'll see how to deal with questions like this in Chapter 7.

Summary

In this chapter, we have looked at queries on a single table. Some of the main points covered are:

- We can return a subset of rows that satisfy a given condition by using a `WHERE` clause. The condition is a Boolean expression, which is a statement that is either true or not true. The condition is applied to each row of the table independently.

- The `SELECT` clause allows us to specify a subset of columns.

- Because the result of a query is a set of rows, we cannot guarantee the order in which the rows will be returned. If we want to display the result in a particular order, we can use the `ORDER BY` clause.

- It is possible to create a view, which essentially stores an SQL command so that you can run it over and over again as the data in the base tables change.

- Tables are likely to have null values (both on purpose and by mistake). Always check how your conditions will apply to null values.

- When you project a subset of columns using an SQL command, the default is to retain duplicate rows in the result. Always think about how you need to deal with the duplicates, and use the keyword `DISTINCT` if you want unique rows.

- The `WHERE` clause considers only one row at a time. Don't use it for queries that require you to look at several rows at once, as in who entered *both* tournaments or who did *not* enter this tournament.

CHAPTER 3

A First Look at Joins

In the previous chapter, we looked at how to retrieve subsets of rows and/or columns from a single table. We saw in Chapter 1 that to keep data accurately in a database, different aspects of the information need to be separated into normalized tables. Most queries will require information from two or more tables. We can combine data from two tables in several different ways depending on the nature of the information we are trying to extract. The most often encountered two-table operation is the join. In Chapter 1 we also introduced two different ways to approach a query: the *process approach* and the *outcome approach*. The first describes how we will combine the tables to achieve the required data, while the second describes what criteria the retrieved data must satisfy.

The Process Approach to Joins

A join enables us to combine related data from two tables. The example we will start with uses the Member and Type tables in order to find the membership fees for each member of the golf club. The first step in carrying out a join is an operation called a Cartesian product.

Cartesian Product

A *Cartesian product* is the most versatile operation between two tables because it can be applied to any two tables of any shape. Having said that, it rarely produces particularly useful information on its own, so its main claim to fame is as the first step of a join.

A Cartesian product is a bit like putting two tables side by side. Let's have a look at the two tables in Figure 3-1: an abbreviated Member table and the Type table.

© Clare Churcher 2016
C. Churcher, *Beginning SQL Queries*, DOI 10.1007/978-1-4842-1955-3_3

MemberID ▾	LastName ▾	FirstName ▾	MemberType ▾
118	McKenzie	Melissa	Junior
138	Stone	Michael	Senior
153	Nolan	Brenda	Senior
176	Branch	Helen	Social
178	Beck	Sarah	Social
228	Burton	Sandra	Junior
235	Cooper	William	Senior
239	Spence	Thomas	Senior
258	Olson	Barbara	Senior
286	Pollard	Robert	Junior

Type ▾	Fee ▾
Associate	60
Junior	150
Senior	300
Social	50

a) (Abbreviated) Member table   b) Type table

Figure 3-1. *Two permanent tables in the database*

The virtual table resulting from the Cartesian product will have a column for each column in the two contributing tables. The rows in the resulting table consist of every combination of rows from the original tables. Figure 3-2 shows the first few rows of the Cartesian product.

	From Member table			From Type table	
MemberID ▾	LastName ▾	FirstName ▾	MemberType ▾	Type ▾	Fee ▾
118	McKenzie	Melissa	Junior	Associate	60
118	McKenzie	Melissa	Junior	Junior	150
118	McKenzie	Melissa	Junior	Senior	300
118	McKenzie	Melissa	Junior	Social	50
138	Stone	Michael	Senior	Associate	60
138	Stone	Michael	Senior	Junior	150
138	Stone	Michael	Senior	Senior	300
138	Stone	Michael	Senior	Social	50
153	Nolan	Brenda	Senior	Associate	60
153	Nolan	Brenda	Senior	Junior	150
153	Nolan	Brenda	Senior	Senior	300
153	Nolan	Brenda	Senior	Social	50

Figure 3-2. *First few rows of the Cartesian product between Member and Type tables*

We have the four columns from the Member table and the two columns from the Type table, which gives us six columns total. Each row from the Member table appears in the resulting table alongside each row from the Type table. We have Melissa McKenzie appearing on four rows – once with each of the four rows in the Type table (Associate, Junior, Senior, Social). The total number of rows will be the number of rows in each table multiplied together; in other words, for this cut-down Member table, we have 10 rows times 4 rows (from Type), giving a total of 40 rows. Cartesian products can produce very, very large result tables, which is why they don't give us much useful information on their own.

A Cartesian product operation is represented in SQL by CROSS JOIN. The SQL to retrieve the data shown in Figure 3-2 is:

```
SELECT *
FROM Member m CROSS JOIN Type t;
```

Not all versions of SQL support the same keywords and phrases (e.g., Microsoft Access 2013 does not support the CROSS JOIN key phrase). In 1992, keywords representing some relational algebra operations (such as CROSS JOIN) were added to the SQL standard,[1] and there have been a number of updates since then. However, not all vendors incorporate all parts of the standard, and other vendors provide additional functionality. Later in the chapter we will look at the outcome approach to provide equivalent ways of expressing queries that will work when the relational algebra operation keywords are not available.

Inner Join

If you look at the table in Figure 3-2, you can see that most of the rows are quite meaningless. For example, the first, third, and fourth rows have the junior member Melissa McKenzie alongside information about the associate, senior, and social membership types. It is difficult to see how these rows will ever be useful. However, the second row, where the member types from each table match, is useful because it allows us to see what fee Melissa pays. If we take just the subset of rows where the value in the MemberType column matches the value in the Type column, then we have useful information about the fees for each of our members. Figure 3-3 shows the rows we would like to retain.

---

[1] International Organization for Standardization. *Information technology — Database languages — SQL*. ISO, Geneva, Switzerland, 1992. ISO/IEC 9075:1992.

MemberID ▾	LastName ▾	FirstName ▾	MemberType ▾	Type ▾	Fee ▾
118	McKenzie	Melissa	Junior	Associate	60
118	McKenzie	Melissa	Junior	Junior	150
118	McKenzie	Melissa	Junior	Senior	300
118	McKenzie	Melissa	Junior	Social	50
138	Stone	Michael	Senior	Associate	60
138	Stone	Michael	Senior	Junior	150
138	Stone	Michael	Senior	Senior	300
138	Stone	Michael	Senior	Social	50
153	Nolan	Brenda	Senior	Associate	60
153	Nolan	Brenda	Senior	Junior	150
153	Nolan	Brenda	Senior	Senior	300
153	Nolan	Brenda	Senior	Social	50

Select rows where these two
columns have the same value

Figure 3-3. *Cartesian product followed by selecting a subset of rows*

The operation shown in Figure 3-3 (a Cartesian product followed by selecting a subset of rows) is known as an *inner join* (often just called a *join*). The condition we use to select the rows is known as the *join condition.* The SQL for the inner join in Figure 3-3 is:

```
SELECT *
FROM Member m INNER JOIN Type t ON m.MemberType = t.Type;
```

The keyword INNER JOIN is used, and we can see the condition for selecting the rows after the keyword ON. Once again, you may find that some versions of SQL do not support the phrase INNER JOIN; however, we will see other ways to express the query later in this chapter.

The two columns that we are comparing (MemberType and Type) must be *join compatible.* Formally, this means they must both come from the same *domain* or set of possible values. In practical terms, join compatibility usually means that the columns in each of the tables have the same data type. For example, both columns will be integers or both dates. Different database products may interpret join compatibility differently. Some might let you join on a float (number with a decimal point) in one table and an integer in another. Some may be fussy about whether text fields are the same length (for example CHAR(10) or CHAR(15)), and others may not. I recommend you don't try to join on fields with different types unless you are very clear what your particular product does. The best strategy, as always, is to think carefully when you design your tables. Those attributes that are likely to be joined should have the same types.

Outcome Approach to Joins

Let's take a look at joins with the outcome approach. Rather than look at how we will combine the tables, we will look at what criteria the retrieved rows must meet.

Let's start with the Cartesian product: we want a set of rows made up of combinations of rows from each of the contributing tables. Figure 3-4 shows how we can envisage this. We are looking at two tables, so we need two fingers to keep track of the rows. Finger m looks at each row of the Member table in turn. Currently it is pointing at row 3. For each row in the Member table, finger t will point to each row in the Type table. For the Cartesian product we retain every combination of the rows. In terms of Figure 3-4 the Cartesian product can be expressed in natural language as:

I'll write out all the attributes from row m and all the attributes from row t so long as m comes from the Member table and t comes from the Type table.

MemberID ▾	LastName ▾	FirstName ▾	MemberType ▾
118	McKenzie	Melissa	Junior
138	Stone	Michael	Senior
m ☞ 153	Nolan	Brenda	Senior
176	Branch	Helen	Social
178	Beck	Sarah	Social
228	Burton	Sandra	Junior
235	Cooper	William	Senior
239	Spence	Thomas	Senior
258	Olson	Barbara	Senior
286	Pollard	Robert	Junior

Member table

Type ▾	Fee ▾
t ☞ Associate	60
Junior	150
Senior	300
Social	50

Type table

Figure 3-4. *Row variables m and t point to each row in the Member and Types tables, respectively*

The SQL for the query represented in Figure 3-4 and that results in the output shown in Figure 3-2 is:

```
SELECT *
FROM Member m, Type t;
```

The preceding statement will return the same rows as the expression we had previously used that used the CROSS JOIN phrase.

For a join we have the extra condition that we want to retrieve only those combinations of rows where the membership type from each table is the same. We can express this in natural language as:

I'll write out all the attributes from row m and all the attributes from row t so long as m comes from the Member table and t comes from the Type table and m.MemberType = t.Type.

The pair of rows depicted in Figure 3-5 satisfies that condition and so will be retrieved. If m stays where it is and t moves down a row, then the condition will no longer be satisfied and the new combination will not be included.

MemberID ▾	LastName ▾	FirstName ▾	MemberType ▾
118	McKenzie	Melissa	Junior
138	Stone	Michael	Senior
m ☞ 153	Nolan	Brenda	Senior
176	Branch	Helen	Social
178	Beck	Sarah	Social
228	Burton	Sandra	Junior
235	Cooper	William	Senior
239	Spence	Thomas	Senior
258	Olson	Barbara	Senior
286	Pollard	Robert	Junior

Type ▾	Fee ▾
Associate	60
Junior	150
t ☞ Senior	300
Social	50

Member table   Type table

Figure 3-5. *Rows will be retrieved where m.MemberType = t.Type*

We can translate the query depicted in Figure 3-5 into SQL as follows:

```
SELECT *
FROM Member m, Type t
WHERE m.MemberType = t.Type;
```

If we look carefully at the preceding statement we can see that the first two lines represent the Cartesian product, and the WHERE clause in last line is selecting a subset of the rows where the membership types are the same. This was how we defined an inner join in the previous section. The preceding statement will produce the same rows as our previous statement for an inner join, seen again here:

```
SELECT *
FROM Member m INNER JOIN Type t ON m.MemberType = t.Type;
```

The first statement says what the rows to be retrieved are like (outcome approach) and the second expresses what operation we should use to retrieve those rows (process approach). Which one you use does not matter–it just depends on how you find yourself thinking about the query. Sometimes there is a possibility that the way you express the query may affect the performance, and we will talk about this more in Chapter 9. Actually, most database products are pretty smart at optimizing, or finding the quickest way to perform a query, regardless of how you express it. For example, in SQL Server the two expressions for the join shown are carried out in the same way. In fact, in SQL Server 2013, if you type the code in the first statement into the default interface for creating a view, it will be replaced by the code using the INNER JOIN phrase.

Extending Join Queries

Now that we have added joins to our arsenal of operations, we can perform numerous types of queries. Because the result of a join (as with any operation) is another table, we can then join that result to a third table (and then another) and then select and project rows and columns to achieve the required result.

Let's look at an example using the tables in Figure 3-6. The Entry table uses two foreign keys (MemberID and TourID) to maintain information about which members have entered the different tournaments. The first line in the Entry table says that member 118 entered tournament 24 in 2014. If we require any additional information (say, the name of a member or name of a tournament), we need to use the foreign keys to find the appropriate rows in the Member and Tournament tables, respectively.

MemberID ▾	LastName ▾	FirstName ▾
118	McKenzie	Melissa
138	Stone	Michael
153	Nolan	Brenda
176	Branch	Helen
178	Beck	Sarah
228	Burton	Sandra
235	Cooper	William
239	Spence	Thomas
258	Olson	Barbara
286	Pollard	Robert
290	Sexton	Thomas
323	Wilcox	Daniel
331	Schmidt	Thomas
332	Bridges	Deborah
339	Young	Betty
414	Gilmore	Jane
415	Taylor	William
461	Reed	Robert
469	Willis	Carolyn
487	Kent	Susan

MemberID ▾	TourID ▾	Year ▾
118	24	2014
228	24	2015
228	25	2015
228	36	2015
235	38	2013
235	38	2015
235	40	2014
235	40	2015
239	25	2015
239	40	2013
258	24	2014
258	38	2014
286	24	2013
286	24	2014
286	24	2015
415	24	2015
415	25	2013
415	36	2014
415	36	2015
415	38	2013
415	38	2015
415	40	2013
415	40	2014
415	40	2015

TourID ▾	TourName ▾
24	Leeston
25	Kaiapoi
36	WestCoast
38	Canterbury
40	Otago

a) Member (Some columns)   b) Entry   c) Tournament

Figure 3-6. *Permanent tables in the club database*

Let's find the names of everyone who entered the Leeston tournament in 2014. I'll describe two different approaches, and you will probably find that one appeals to you more than the other.

A Process Approach

We are starting with three tables, so we need some operation that combines data from more than one table. We can join the Member table to the Entry table and the result to the Tournament table, as shown in Figure 3-7.

Member joined with Entry on
m.MemberID = e.MemberID

Join result to Tournament on
e.TourID = t.TourID

m.MemberID ▾	LastName ▾	FirstName ▾	e.MemberID ▾	e.TourID ▾	Year ▾	t.TourID ▾	TourName ▾	TourType ▾
118	McKenzie	Melissa	118	24	2014	24	Leeston	Social
228	Burton	Sandra	228	24	2015	24	Leeston	Social
228	Burton	Sandra	228	25	2015	25	Kaiapoi	Social
228	Burton	Sandra	228	36	2015	36	WestCoast	Open
235	Cooper	William	235	38	2013	38	Canterbury	Open
235	Cooper	William	235	38	2015	38	Canterbury	Open
235	Cooper	William	235	40	2014	40	Otago	Open
235	Cooper	William	235	40	2015	40	Otago	Open
239	Spence	Thomas	239	25	2015	25	Kaiapoi	Social
239	Spence	Thomas	239	40	2013	40	Otago	Open
258	Olson	Barbara	258	24	2014	24	Leeston	Social
258	Olson	Barbara	258	38	2014	38	Canterbury	Open
286	Pollard	Robert	286	24	2013	24	Leeston	Social
286	Pollard	Robert	286	24	2014	24	Leeston	Social
286	Pollard	Robert	286	24	2015	24	Leeston	Social
415	Taylor	William	415	24	2015	24	Leeston	Social
415	Taylor	William	415	25	2013	25	Kaiapoi	Social
415	Taylor	William	415	36	2014	36	WestCoast	Open
415	Taylor	William	415	36	2015	36	WestCoast	Open
415	Taylor	William	415	38	2013	38	Canterbury	Open
415	Taylor	William	415	38	2015	38	Canterbury	Open
415	Taylor	William	415	40	2013	40	Otago	Open
415	Taylor	William	415	40	2014	40	Otago	Open
415	Taylor	William	415	40	2015	40	Otago	Open

From Member table (m)   From Entry table (e)   From Tournament table (t)

Figure 3-7. *Joining the Member, Entry, and Tournament tables*

The join condition for the first join between the Member and Entry tables is that m.MemberID = e.MemberID as shown by the rectangular boxes in Figure 3-7. For the second join between the result of the first join and the Tournament table, the condition is that e.TourID = t.TourID as shown by the circles. It will not make any difference if we choose to do the join between Entry and Tournament first and then join the result to Member.

The SQL to carry out the two joins is:

```
SELECT *
FROM (Member m INNER JOIN Entry e ON m.MemberID = e.MemberID)
     INNER JOIN Tournament t ON e.TourID = t.TourID;
```

The virtual table resulting from the two joins in this query has all the information we require to answer our question. We just need to select the rows satisfying the conditions about the year and tournament name by adding a WHERE clause, and then project the name attributes by specifying them in the SELECT clause. The complete SQL query to return the names of everyone who entered the Leeston tournament in 2014 is:

```
SELECT LastName, FirstName
FROM (Member m INNER JOIN Entry e ON m.MemberID = e.MemberID)
     INNER JOIN Tournament t ON e.TourID = t.TourID
WHERE TourName = 'Leeston'
AND Year = 2014;
```

Order of Operations

In the description in the previous section, we joined all the tables first and then selected the appropriate rows and columns. The result of the join is an intermediate table (as in Figure 3-7) that is potentially extremely large if there are lots of members and tournaments. We could have done the operations in a different order. We could have first selected just the Leeston tournament from the Tournament table and the 2014 tournaments from the Entry tables, as shown in Figure 3-8. Joining these two smaller tables with each other and then joining that result with Member would result in a much smaller intermediate table.

Select 2014 entries

MemberID ▾	TourID ▾	Year ▾
118	24	2014
235	40	2014
258	24	2014
258	38	2014
286	24	2014
415	36	2014
415	40	2014

Select the Leeston tournament

TourID ▾	TourName ▾	TourType ▾
24	Leeston	Social

Figure 3-8. *Selecting rows from the Entry and Tournament tables before joining them*

So, should we worry about the order of the operations? The answer is "yes" – the order of operations makes a huge difference – but if you are using SQL, then it is not your problem to worry about. The SQL statement is always going to be the same, but with the tables possibly in a different order. The SQL statement is sent to the engine of whatever database program you are using, and the query will be *optimized*. This means the database program figures out the best order to do things. Some products do this extremely well, others not so well. Many products have analyzer tools that will let you see in what order things are being done. For many queries, writing your SQL differently doesn't make much difference, but you can make things more efficient by providing indexes for your tables. We will look at these issues more closely in Chapter 9.

An Outcome Approach

The reason that the way we write our SQL statements often doesn't affect the efficiency of a query is that SQL is fundamentally based on relational calculus, which describes the criteria the retrieved rows must meet. The original SQL standards did not even have keywords like INNER JOIN. SQL statements without these keywords describe *what* the retrieved rows should be like, so they do not have anything to say about *how*. Let's look at an outcome approach to finding the names of members who entered Leeston tournaments in 2014.

We want to retrieve just some names from the Member table. Forget joins, and think about how you would know whether a particular name should be retrieved if you were shown the three tables and knew nothing about databases or foreign keys or joins or anything. Imagine a finger m tracing down the table, as in Figure 3-9.

MemberID	LastName	FirstName		MemberID	TourID	Year		TourID	TourName
118	McKenzie	Melissa		118	24	2014	t ☞	24	Leeston
138	Stone	Michael		228	24	2015		25	Kaiapoi
153	Nolan	Brenda		228	25	2015		36	WestCoast
176	Branch	Helen		228	36	2015		38	Canterbury
178	Beck	Sarah		235	38	2013		40	Otago
228	Burton	Sandra		235	38	2015			
235	Cooper	William		235	40	2014			
239	Spence	Thomas		235	40	2015			
m ☞ 258	Olson	Barbara		239	25	2015			
286	Pollard	Robert		239	40	2013			
290	Sexton	Thomas	e ☞	258	24	2014			
323	Wilcox	Daniel		258	38	2014			
331	Schmidt	Thomas		286	24	2013			
332	Bridges	Deborah		286	24	2014			
339	Young	Betty		286	24	2015			
414	Gilmore	Jane		415	24	2015			
415	Taylor	William		415	25	2013			
461	Reed	Robert		415	36	2014			
469	Willis	Carolyn		415	36	2015			
487	Kent	Susan		415	38	2013			
				415	38	2015			
				415	40	2013			
				415	40	2014			
				415	40	2015			

a) Member (Some columns)	b) Entry	c) Tournament

Figure 3-9. *Using row variables to describe the rows that satisfy the query conditions*

Do we want to write out Barbara Olson, the name to which m is currently pointing? How would we know? Well, first we have to find a row with her ID (235) in the Entry table for the year 2014 such as the one where finger e is pointing. Then we have to find a row with that tournament ID (24) in the Tournament table and check it is a Leeston tournament. Looking at Figure 3-9, we see that the rows where the three fingers are pointing give us enough information to know that Barbara Olson did indeed enter a Leeston tournament in 2014. This set of conditions describes *what* a row in the result table should be like.

Now let's write that last paragraph a bit more succinctly. Read the following sentence with reference to the rows denoted in Figure 3-9:

> *I'll write out the names from row* m, *where* m *comes from the* Member *table, if there is a row* e *in the* Entry *table where* m.MemberID *is the same as* e.MemberID *and* e.Year *is 2014 and there also exists a row* t *in the* Tournament *table where* e.TourID *is the same as* t.TourId *and* t.TourName *has the value "Leeston."*

The SQL reflects the preceding paragraph. Look carefully at the following statement with reference to Figure 3-9:

```
SELECT m.LastName, m.FirstName
FROM Member m, Entry e, Tournament t
WHERE m.MemberID = e.MemberID
      AND e.TourID = t.TourID
      AND t.TourName = 'Leeston' AND e.Year = 2014;
```

You can see how the SQL statement describes *what* a retrieved row should be like. If you look carefully at the statement, you can also spot the operations. The second line (the FROM clause) is a big Cartesian product, the next two lines are the join conditions (which would result in a table like the one in Figure 3-7), the final line selects the rows with the appropriate year and tournament name, and the SELECT clause line tells us to project just the names.

The SQL preceding statement is equivalent to the one using the INNER JOIN keywords. They will both return the same set of rows: one reflects the underlying process of *how*, and the other reflects the underlying outcome of *what*.

Expressing Joins Through Diagrammatic Interfaces

This book is about queries in SQL, but most database products also provide a diagrammatic interface to express queries. Just for completeness, I'll show you what a typical diagrammatic interface looks like for retrieving the names of members who entered the Leeston tournament in 2014.

Figure 3-10 shows the Microsoft Access interface, but most products have something very similar. The tables are represented by the rectangles in the top section with the lines showing the joins between them. The columns to be retrieved have a checkmark (√) in the row marked Show, and the conditions for selecting a particular row are shown for the relevant fields in the row marked Criteria.

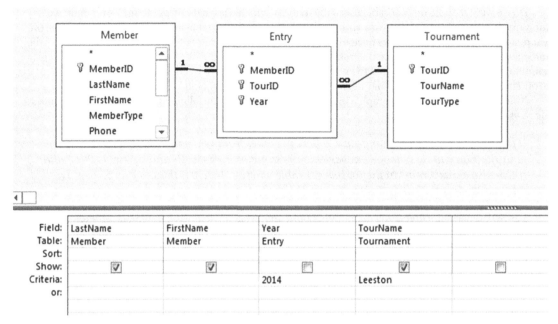

Figure 3-10. *Microsoft Access digrammatic interface for the query to find names of members entering the Leeston tournament in 2014*

Other Types of Joins

The joins we have been looking at in this chapter are *equi-joins*. An equi-join is one where the join condition has an equals operator, as in m.MemberID = e.MemberID. This is the most common type of condition, but you can have different operators. A join is just a Cartesian product followed by selecting a subset of rows, and the select condition can consist of different comparison operators (for example, <> or >) and also logical operators (for example, AND or NOT). These sorts of joins don't turn up all that often.

You might also come across a *natural join*. A natural join assumes that you will be joining on columns that have the same name in both tables. The join condition is that the values in the two columns with the same name are equal, and one of those columns will be removed from the result. For example:

```
SELECT * FROM
Member NATURAL JOIN Entry;
```

This would produce almost the same output as:

```
SELECT * FROM
Member m INNER JOIN Entry m ON m.MemberID = e.MemberID;
```

In the natural join statement, the join condition is implicitly assumed to be equality between the two attributes with the same name, MemberID. The only difference between the two queries is that for the natural join only one of the MemberID columns will be returned. Oracle supports natural joins but SQL Server and Access do not.

Outer Joins

One type of join that you will use a great deal and that is important to understand is the *outer join*. The best way to understand an outer join is to see where they are useful. Have a look at the (modified) Member and Type tables in Figure 3-11.

MemberID ▾	LastName ▾	FirstName ▾	MemberType ▾
118	McKenzie	Melissa	Junior
138	Stone	Michael	Senior
153	Nolan	Brenda	Senior
176	Branch	Helen	Social
178	Beck	Sarah	
228	Burton	Sandra	Junior
235	Cooper	William	Senior
239	Spence	Thomas	Senior
258	Olson	Barbara	Senior

Type ▾	Fee ▾
Associate	60
Junior	150
Senior	300
Social	50

(modified) Member table   Type table

Figure 3-11. *Member and Type tables*

You might want to produce different lists from the Member table, such as numbers and names, names and membership types, and so on. In these lists you expect to see all the members (for the table in Figure 3-11, that would be nine rows). Then you might think that as well as seeing the numbers and names in your member list, you will also include the membership fee. You join the two tables (with the condition MemberType = Type) and find that you "lose" one of your members – Sarah Beck (see Figure 3-12).

MemberID ▾	LastName ▾	FirstName ▾	MemberType ▾	Type ▾	Fee ▾
118	McKenzie	Melissa	Junior	Junior	150
138	Stone	Michael	Senior	Senior	300
153	Nolan	Brenda	Senior	Senior	300
176	Branch	Helen	Social	Social	50
228	Burton	Sandra	Junior	Junior	150
235	Cooper	William	Senior	Senior	300
239	Spence	Thomas	Senior	Senior	300
258	Olson	Barbara	Senior	Senior	300

Figure 3-12. *Inner join between Member and Type, and we "lose" Sarah Beck*

The reason is that Sarah has no value for MemberType in the Member table. Let's look at the Cartesian product, which is the first step for doing a join. Figure 3-13 shows those rows of the Cartesian product that include Sarah.

MemberID ▾	LastName ▾	FirstName ▾	MemberType ▾	Type ▾	Fee ▾
176	Branch	Helen	Social	Associate	60
176	Branch	Helen	Social	Junior	150
176	Branch	Helen	Social	Senior	300
176	Branch	Helen	Social	Social	50
178	Beck	Sarah		Associate	60
178	Beck	Sarah		Junior	150
178	Beck	Sarah		Senior	300
178	Beck	Sarah		Social	50
228	Burton	Sandra	Junior	Associate	60
228	Burton	Sandra	Junior	Junior	150
228	Burton	Sandra	Junior	Senior	300
228	Burton	Sandra	Junior	Social	50

Figure 3-13. *Part of the Cartesian product between the Member and Type tables*

Having done the Cartesian product, we now need to do the final part of our join operation, which is to apply the condition (MemberType = Type). As you can see in Figure 3-13, there is no row for Sarah Beck that satisfies this condition because she has a null or empty value in MemberType.

Consider the following two natural-language questions: "Get me the fees for members" and "Get me all member information including fees." The first one has an implication of "Just get me the members who have fees," while the second has more of a feel of "Get me all the members and include the fees for those who have them." One of the biggest difficulties in writing queries is trying to decide exactly what it is you want. It is even more difficult if you are trying to understand what someone else is asking for!

Let's say that what we actually want is a list of all our members, and where we can find the fee information, we'd like to include that. In this case, we want to see Sarah Beck included in the result, but with no fee displayed. That is what an outer join does. Outer joins can come in three types: left, right, and full outer joins. A left outer join retrieves all the rows from the left table including those with a null value in the join field, as shown in Figure 3-14. We see that as well as all the rows from the inner join (Figure 3-12), we also have a row from the Member table for Sarah, who had a null for the join field MemberType. The fields in that row that would have come from the right-hand table (Type and Fee) have null values.

MemberID ▾	LastName ▾	FirstName ▾	MemberType ▾	Type ▾	Fee ▾
118	McKenzie	Melissa	Junior	Junior	150
138	Stone	Michael	Senior	Senior	300
153	Nolan	Brenda	Senior	Senior	300
176	Branch	Helen	Social	Social	50
178	Beck	Sarah			
228	Burton	Sandra	Junior	Junior	150
235	Cooper	William	Senior	Senior	300
239	Spence	Thomas	Senior	Senior	300
258	Olson	Barbara	Senior	Senior	300

Figure 3-14. *Result of left outer join between Member and Type tables*

The SQL for the outer join depicted in Figure 3-14 is similar to an inner join, but the key phrase INNER JOIN is replaced with LEFT OUTER JOIN (or in some applications simply LEFT JOIN):

```
SELECT *
FROM Member m LEFT OUTER JOIN Type t ON m.MemberType = t.Type;
```

You might quite reasonably say that we wouldn't have needed an outer join if all the members had a value for the MemberType field (as they probably should). That may be true for this case – but remember my cautions in Chapter 2 about assuming that fields that *should* have data *will* have data. In other situations, the data in the join field may be quite legitimately empty. We will see in later chapters queries like "List all members and the names of their coaches – if they have one." "Losing" rows because you have used an inner join when you should have used an outer join is a very common problem and is sometimes quite hard to spot.

What about right and full outer joins? Left and right outer joins are the same and just depend on which order you put the tables in the join statement. The following SQL statement will return the same information as displayed in Figure 3-14, although the columns may be presented in a different order:

```
SELECT *
FROM Type t RIGHT OUTER JOIN Member m ON m.MemberType = t.Type;
```

We have simply swapped the order of the tables in the join statement. Any rows with a null in the join field of the right table (Member) will be included.

A full outer join will retain rows with a null in the join field in either table. The SQL for the full outer join is shown here and will result in the table seen in Figure 3-15:

```
SELECT *
FROM Member m FULL OUTER JOIN Type t ON m.MemberType = t.Type;
```

MemberID ▾	LastName ▾	FirstName ▾	MemberType ▾	Type ▾	Fee ▾
				Associate	60
118	McKenzie	Melissa	Junior	Junior	150
138	Stone	Michael	Senior	Senior	300
153	Nolan	Brenda	Senior	Senior	300
176	Branch	Helen	Social	Social	50
178	Beck	Sarah			
228	Burton	Sandra	Junior	Junior	150
235	Cooper	William	Senior	Senior	300
239	Spence	Thomas	Senior	Senior	300
258	Olson	Barbara	Senior	Senior	300

Figure 3-15. *Result of a full outer join between Member and Type tables*

We have a row for Sarah Beck padded with mull values for the missing columns from the Type table. We also have the first row, which shows us the information about the Associate membership type even though there are no rows in the Member table with Associate as a member type. In this row, each missing value from the Member table is replaced with a null.

Not all implementations of SQL have a full outer join implemented explicitly. Access 2013 doesn't. However, there are always alternative ways in SQL to retrieve the information you require. In Chapter 7 I'll show you how to get the equivalent of a full outer join by using a union operator between a left and right outer join (which is what I had to do to get the screen shot in Figure 3-15!).

Summary

A Cartesian product combines two tables. The resulting table has a column for each column in the two tables, and there is a row for every combination of rows from the contributing tables. The SQL for a Cartesian product reflecting the process approach is:

```
SELECT *
FROM <table1> CROSS JOIN <table2>;
```

The SQL for an inner join reflecting the outcome approach is:

```
SELECT *
FROM <table1>,<table2>;
```

An inner join starts with a Cartesian product, and then a join condition determines which combinations of rows from the two contributing tables will be retained.

The SQL for an inner join reflecting the process approach is:

```
SELECT *
FROM <table1> INNER JOIN <table2>
ON <join condition>;
```

The SQL for an inner join reflecting the outcome approach is:

```
SELECT *...
FROM <table1>, <table2>
WHERE <join condition>;
```

If one (or both) of the tables has rows with a null in the field involved in the join condition, then that row will not appear in the result for an inner join. If that row is required, you can use outer joins.

The SQL for an outer join, which will retain all the rows in the left-hand table including those with a null in the join field, is:

```
SELECT *
FROM  <table1> LEFT OUTER JOIN <table2>
ON <join condition>;
```

Similar expressions exist for right outer joins and full outer joins.

CHAPTER 4

Subqueries

In the previous chapters, we looked at retrieving a subset of rows and columns from a single table, and we also looked at how Cartesian products and joins can be used to retrieve data from two or more tables. In many of the examples it was possible to construct quite different SQL queries to produce the same result. Depending on the context or the problem you will probably find that one approach will feel more natural.

As queries become more complicated, we might find that we can think of SQL expressions for small parts of a query but not for the whole lot in one go. It is possible to return data from a query and then refer to that data with another query – all in the one SQL statement. This idea of a query within a query is very powerful. You will hear the concept referred to as a query and subquery or inner and outer queries or nested queries.

In this chapter, we will look at subqueries and two new SQL keywords, EXISTS and IN. We will see how to use subqueries as an alternative way to approach some of the queries we have already done and also how nesting will open up other possibilities.

IN Keyword

The IN keyword allows us to select rows from a table, where the condition allows an attribute to have one of several values. For example, if we wanted to retrieve the member IDs from the rows in our Entry table for tournaments with ID 36, 38, or 40, we could do this with a Boolean OR operator as in the following query:

```
SELECT e.MemberID
FROM Entry e
WHERE e.TourID = 36 OR e.TourID = 38 OR e.TourID = 40;
```

Clearly, statements of this type will start to become unwieldy as the number of possible options grows. Using the IN keyword, we can construct a more compact statement where the set of possible values are enclosed in parentheses and separated by commas. In the following query, each row of Entry is investigated, and if TourID is one of the values in the parentheses, then the WHERE condition is true, and that row will be returned:

```
SELECT e.MemberID
FROM Entry e
WHERE e.TourID IN (36, 38, 40);
```

© Clare Churcher 2016

C. Churcher, *Beginning SQL Queries*, DOI 10.1007/978-1-4842-1955-3_4

It is possible to combine IN with the logical operator NOT. However, you need to be very careful. Consider the following query:

```
SELECT e.MemberID
FROM Entry e
WHERE e.TourID NOT IN (36, 38, 40);
```

The preceding query will return the IDs of members who have entered any tournament that is not in the list. Be aware though that those members may have entered one of the tournaments in the list as well. We will look at how to accurately answer questions such as "who has not entered these tournaments" later in this chapter.

Using IN with Subqueries

The real usefulness of the IN keyword is that we can use another SQL statement to generate the set of values. For example, the reason that someone may have been interested in tournaments 36, 38, and 40 might have been because they are the current Open tournaments. Rather than list the Open tournaments individually, we can use another SQL query to generate the set of values we require. The list will be reconstructed each time the query is run so that the set of Open tournaments will remain current as the data changes.

Let's look at a specific example of using a query to generate the set of values for the IN clause. I've reproduced a few of the columns of the Member table along with the Entry and Tournament tables in Figure 4-1.

MemberID ▾	LastName ▾	FirstName ▾
118	McKenzie	Melissa
138	Stone	Michael
153	Nolan	Brenda
176	Branch	Helen
178	Beck	Sarah
228	Burton	Sandra
235	Cooper	William
239	Spence	Thomas
258	Olson	Barbara
286	Pollard	Robert
290	Sexton	Thomas
323	Wilcox	Daniel
331	Schmidt	Thomas
332	Bridges	Deborah
339	Young	Betty
414	Gilmore	Jane
415	Taylor	William
461	Reed	Robert
469	Willis	Carolyn
487	Kent	Susan

(Some columns) Member

MemberID ▾	TourID ▾	Year ▾
118	24	2014
228	24	2015
228	25	2015
228	36	2015
235	38	2013
235	38	2015
235	40	2014
235	40	2015
239	25	2015
239	40	2013
258	24	2014
258	38	2014
286	24	2013
286	24	2014
286	24	2015
415	24	2015
415	25	2013
415	36	2014
415	36	2015
415	38	2013
415	38	2015
415	40	2013
415	40	2014
415	40	2015

Entry

TourID ▾	TourName ▾	TourType ▾
24	Leeston	Social
25	Kaiapoi	Social
36	WestCoast	Open
38	Canterbury	Open
40	Otago	Open

Tournament

Figure 4-1. *Member, Entry, and Tournament tables*

The query to generate the set of IDs for the Open tournaments is:

```
SELECT t.TourID
FROM Tournament t
WHERE t.TourType = 'Open';
```

Now we can replace the list of explicit values (36, 38, 40) in the previous queries with the preceding SQL statement:

```
SELECT e.MemberID
FROM Entry e
WHERE e.TourID IN (
    SELECT t.TourID
    FROM Tournament t
    WHERE t.TourType = 'Open');
```

The SELECT statement inside the parentheses is sometimes referred to as a *subquery*. To work correctly with the IN keyword, the inner part of the query must return a list of single values. I have indented it only to make it easier to read (SQL will ignore the added whitespace). You can understand a nested query by reading it from the "inside out." The inside SELECT statement retrieves the set of required tournament IDs from the Tournament table, and then the outside SELECT finds us all the entries from the Entry table for tournaments IN that set.

To aid in understanding, it is possible to add comments to SQL statements. In the statement that follows the line beginning with -- is a comment and will be ignored. It is also possible to use /* and */ around a block of more than one line of code.

```
SELECT e.MemberID
FROM Entry e
WHERE e.TourID IN (
    -- Subquery returns IDs of Open tournaments
    SELECT t.TourID
    FROM Tournament t
    WHERE t.TourType = 'Open');
```

Have another look at the tables in Figure 4-1. How else might we have retrieved entries for Open tournaments? We carried out similar queries in the previous chapter using a join. We can join the two tables, Entry and Tournament, on their common fields TourID, select just those rows that are for Open tournaments, and then project the MemberID column. See the following:

```
SELECT e.MemberID
FROM Entry e INNER JOIN Tournament t ON e.TourID = t.TourID
WHERE t.TourType = 'Open';
```

The SQL statements with and without the subquery retrieve the same information. As I've said a number of times, there are often several different ways to write a query in SQL. The more methods you are familiar with, the more likely you will be able to find a way to express a complicated query.

Being Careful with NOT and <>

As well as asking a question such as "What are the IDs of members who have entered an Open tournament?" it is just as likely that we might want to know "What are the IDs of members who have *not* entered an Open tournament?" They sound very similar, but once we start using negatives in our questions, we have to be very careful about what we really mean. In Chapter 7, we will investigate constructing queries using set operations, but to keep this chapter complete, I'll talk about how negatives impact the use of subqueries in particular.

In the previous section we constructed two SQL statements for retrieving member IDs for members who have entered an Open tournament. One used a subquery and one a join. To find who has not entered an Open tournament, one might attempt changing IN to NOT IN in the subquery example, as follows:

```
SELECT e.MemberID
FROM Entry e
WHERE e.TourID NOT IN
    (SELECT t.TourID
    FROM Tournament t
    WHERE t.TourType = 'Open');
```

In the join example there is a temptation to amend t.TourType = 'Open' to t.TourType <> 'Open':

```
SELECT e.MemberID
FROM Entry e INNER JOIN Tournament t ON e.TourID = t.TourID
WHERE t.TourType <>'Open';
```

Carefully think about which rows will be returned by these two queries. They in fact both return the same set of rows, but those rows may include members who have entered an Open tournament as well as those who have not.

The table in Figure 4-2 shows the result of the inner join between Entry and Tournament. The bottom set of rows are all for Open tournaments, and these will be retrieved by a query that has the condition WHERE t.TourType = 'Open'. The top set of entries is for tournaments other than Open and will be retrieved by the query which has the condition WHERE t.TourType <> 'Open'.

MemberID ▾	e.TourID ▾	Year ▾	t.TourID ▾	TourName ▾	TourType ▾
118	24	2014	24	Leeston	Social
228	24	2015	24	Leeston	Social
258	24	2014	24	Leeston	Social
286	24	2013	24	Leeston	Social
286	24	2014	24	Leeston	Social
286	24	2015	24	Leeston	Social
415	24	2015	24	Leeston	Social
228	25	2015	25	Kaiapoi	Social
239	25	2015	25	Kaiapoi	Social
415	25	2013	25	Kaiapoi	Social
228	36	2015	36	WestCoast	Open
415	36	2014	36	WestCoast	Open
415	36	2015	36	WestCoast	Open
235	38	2013	38	Canterbury	Open
235	38	2015	38	Canterbury	Open
258	38	2014	38	Canterbury	Open
415	38	2013	38	Canterbury	Open
415	38	2015	38	Canterbury	Open
235	40	2014	40	Otago	Open
235	40	2015	40	Otago	Open
239	40	2013	40	Otago	Open
415	40	2013	40	Otago	Open
415	40	2014	40	Otago	Open
415	40	2015	40	Otago	Open

TourType <> 'Open' (top ten rows); TourType = 'Open' (remaining rows)

Figure 4-2. *TourType = 'Open' versus TourType <> 'Open'*

We can see that some members (indicated by circles) appear in both sets. Figure 4-3 is another representation of the information in the table in Figure 4-2 but shows two sets of members rather than entries: the top circle represents those who have entered an Open tournament and the bottom circle those who have entered a tournament that is not an Open tournament. Four members are in both sets.

Open Tournaments

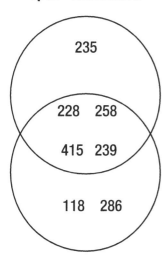

Other Tournaments

Figure 4-3. *Members who have entered Open tournaments, other tournaments, or both*

Now let's return to the original question. Which members have not entered an Open tournament? We have to be careful to differentiate the two sets depicted in Figure 4-4.

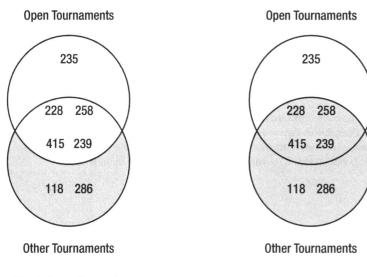

a. Shaded area is people who have not entered an Open tournament

b. Shaded area is people who entered a tournament that is not an Open tournament

Figure 4-4. *It is important to be careful to distinguish the SQL for these two situations*

Figure 4-4a shows the set of people who have not entered any Open tournament. Figure 4-4b shows those members who have entered something other than an Open tournament (but not excluding those who may have entered an Open tournament as well). For example, member 118 has never entered an Open tournament, whereas member 228 has entered Open tournaments as well as other types of tournaments.

The two queries at the beginning of this section will both retrieve the set of members depicted in Figure 4-4b. Member 228 (who has entered an Open tournament) will be returned (because he has entered a tournament that is not an Open tournament). This is not what we want and is a very common mistake.

To decide whether someone has entered an Open competition, we need to find just *one* matching entry. To decide whether someone has *not* entered an Open competition, we need to check *all* the Open entries to make sure that member does not appear.

In terms of our joined tables in Figure 4-2, finding those people who have entered an Open tournament requires a simple WHERE clause: WHERE t.TourType = 'Open'. Remember that each row is inspected independently to decide whether it meets the criteria in a WHERE clause. However, to find people who have *not* entered an Open tournament, we need to investigate *every* row in the table to ensure that there is not an entry for a particular member. This is a much more complex task. In fact, we also need to consider the members who have never entered any tournaments. These members' IDs will not appear in the Entry table at all, so we also have to investigate the Member table to find the complete list.

Finding members who have not entered an Open tournament can be achieved with a process approach using the set operations found in Chapter 7. However, we can also use the outcome approach to construct an accurate query. To do that, we need to first introduce the EXISTS keyword.

EXISTS Keyword

Let's start with a simple question. For example, "What are the names of all members who have ever entered any tournament?" We can start by thinking in terms of which rows of the Member table would satisfy our question. Consider the following sentence and Figure 4-5 together:

I'll write out the names from row m, where m comes from the Member table, if there exists a row e in the Entry table where m.MemberID = e.MemberID.

MemberID ▾	LastName ▾	FirstName
118	McKenzie	Melissa
138	Stone	Michael
153	Nolan	Brenda
176	Branch	Helen
178	Beck	Sarah
228	Burton	Sandra
235	Cooper	William
239	Spence	Thomas
258	Olson	Barbara
286	Pollard	Robert
290	Sexton	Thomas
323	Wilcox	Daniel
331	Schmidt	Thomas
332	Bridges	Deborah
339	Young	Betty
414	Gilmore	Jane
415	Taylor	William
461	Reed	Robert
469	Willis	Carolyn
487	Kent	Susan

m ☞ (235 circled)

Member

MemberID ▾	TourID ▾	Year ▾
118	24	2014
228	24	2015
228	25	2015
228	36	2015
235	38	2013
235	38	2015
235	40	2014
235	40	2015
239	25	2015
239	40	2013
258	24	2014
258	38	2014
286	24	2013
286	24	2014
286	24	2015
415	24	2015
415	25	2013
415	36	2014
415	36	2015

e ☞ (235 circled)

Entry

Figure 4-5. William Cooper has entered a tournament because a matching row exists in the Entry table

We can translate the statement

I'll write out the names from row m, where m comes from the Member table, if there exists a row e in the Entry table where m.MemberID = e.MemberID.

almost directly into SQL with the use of the keyword EXISTS:

```
SELECT m.LastName, m.FirstName
FROM Member m
WHERE EXISTS
    (SELECT * FROM Entry e WHERE e.MemberID = m.MemberID);
```

This is another example of a nested query where we have two SQL SELECT statements, one inside the other. This one is a little different from the simpler example we saw earlier in the chapter. The WHERE condition in the inner query refers to part of the row being considered in the outer query; that is, e.MemberID = m.MemberID. I find the easiest way to interpret these nested queries is with reference to a diagram like Figure 4-5. Variable m is checking each row in the Member table. The inner query is looking for a row in the Entry table with the same value for MemberID as the row under consideration in the Member table. If such a row (or several such rows) EXIST, then we are in business.

For those of you who are thinking that this seems like a complicated way to get a simple result, you are right (partly). The query using the EXISTS clause retrieves the same members as an inner join (on MemberID) between Member and Entry does.

However, what if we want those members who have *not* entered a tournament? This requires only a tiny change to our new SQL query. Instead of looking for members where a matching row in Entry exists, we now want those where a matching row does *not* exist. Adding the word NOT to the previous SQL statements gives us what we require:

```
SELECT m.Lastname, m.FirstName
FROM Member m
WHERE NOT EXISTS
    (SELECT * FROM Entry e WHERE e.MemberID = m.MemberID);
```

The NOT EXISTS construction will look through every row e in the Entry table, checking whether there is a row matching the MemberID of the current row in the Member table. The name of the member will be retrieved only if *no* matching row is found.

Now we have enough ammunition to tackle the query about members who have not entered an Open tournament. Check out Figure 4-6 to decide if William Cooper should be included in the result.

MemberID ▾	LastName ▾	FirstName ▾
118	McKenzie	Melissa
138	Stone	Michael
153	Nolan	Brenda
176	Branch	Helen
178	Beck	Sarah
228	Burton	Sandra
235	Cooper	William
239	Spence	Thomas
258	Olson	Barbara
286	Pollard	Robert
290	Sexton	Thomas
323	Wilcox	Daniel
331	Schmidt	Thomas
332	Bridges	Deborah
339	Young	Betty
414	Gilmore	Jane
415	Taylor	William
461	Reed	Robert
469	Willis	Carolyn
487	Kent	Susan

Member

MemberID ▾	TourID ▾	Year ▾
118	24	2014
228	24	2015
228	25	2015
228	36	2015
235	38	2013
235	38	2015
235	40	2014
235	40	2015
239	25	2015
239	40	2013
258	24	2014
258	38	2014
286	24	2013
286	24	2014
286	24	2015
415	24	2015
415	25	2013
415	36	2014
415	36	2015

Entry

TourID ▾	TourName ▾	TourType ▾
24	Leeston	Social
25	Kaiapoi	Social
36	WestCoast	Open
38	Canterbury	Open
40	Otago	Open

Tournament

Figure 4-6. *There does exist an entry for an Open tournament for William Cooper*

59

The rows indicated in Figure 4-6 show that there does exist an entry for William Cooper, so we will not include him our result.

Now, look at this natural language statement that describes Figure 4-6:

I'll write out the names from row m, where m comes from the Member table, so long as there does not exist (a row e in the Entry table where m.MemberID = e.MemberID along with a row t in the Tournament table where e.TourID = t.TourID and t.TourType = 'Open')

The SQL reflecting the preceding statement is:

```
SELECT m.Lastname, m.FirstName
FROM Member m
WHERE NOT EXISTS
    (SELECT * FROM Entry e, Tournament t
    WHERE m.MemberID = e.MemberID
    AND e.TourID = t.TourID AND t.TourType = 'Open');
```

We will look at the process approach to queries like this one when we cover set operations in Chapter 7.

Different Types of Subqueries

We saw different types of subqueries in the previous sections. It is useful to review some of the options here. The inner part of the nested query can return a single value (e.g, Barbara's handicap), a set of values (e.g., the IDs of Open tournaments), or a set of rows (e.g., entries in Open tournaments). Also, the inner and outer queries can be independent to some extent, or they can be correlated.

Inner Queries Returning a Single Value

Inner queries that return a single value are often useful in the situation where you are simply retrieving a subset of rows. Let's consider the handicaps of our members, as shown in Figure 4-7.

MemberID ▾	LastName ▾	FirstName ▾	Handicap ▾
118	McKenzie	Melissa	30
138	Stone	Michael	30
153	Nolan	Brenda	11
176	Branch	Helen	
178	Beck	Sarah	
228	Burton	Sandra	26
235	Cooper	William	14
239	Spence	Thomas	10
258	Olson	Barbara	16
286	Pollard	Robert	19
290	Sexton	Thomas	26

Figure 4-7. Part of the Member table showing names and handicaps

If we want to find those members with a handicap of less than 16, then this can be done simply with the following SQL:

```
SELECT *
FROM Member m
WHERE m.Handicap < 16;
```

What should we do if we want to find all the members with a handicap less than Barbara Olson's? The preceding query will do that for us, but only if Barbara's handicap of 16 doesn't change. For the query to work for whatever Barbara's current handicap is, we can replace the single value 16 with the result of an inner query:

```
SELECT *
FROM Member m
WHERE Handicap <
    (SELECT Handicap
    FROM Member
    WHERE LastName = 'Olson' AND FirstName = 'Barbara');
```

We need to compare Handicap with a single value. If in a situation like this our inner query returns more than one value (for example, if there were more than one Barbara Olson in the club), then we would get an error when trying to run the query.

An inner query returning a single value is also useful if we want to compare values with an aggregate of some sort. For example, we might want to find all the members who have a handicap less than the average. In this case, we can use the inner query to return the average value:

```
SELECT *
FROM Member m
WHERE m.Handicap <
    (SELECT AVG(Handicap)
    FROM Member);
```

If you take it nice and slow, you can gradually build up quite complicated queries. Say we want to see whether any junior members have a lower handicap than the average for seniors. The inner query has to return the average value handicap for a senior member, and then we want to select all juniors with a handicap less than that. In the SQL statement that follows, both the inner and outer queries have an extra SELECT condition (the inner retrieves just seniors, and the outer retrieves just juniors):

```
SELECT *
FROM Member m
WHERE m.MemberType = 'Junior' AND Handicap < (
    SELECT AVG(Handicap)
    FROM Member
    WHERE MemberType = 'Senior');
```

Inner Queries Returning a Set of Values

This is where we started this chapter. When we use the IN keyword, SQL will expect to find a set of single values. For example, we might ask for rows from the Entry table for members with IDs IN a set of values. In the following statement, the inner query selects the IDs of all senior members, and the outer query returns the entries for those members:

```
SELECT *
FROM Entry e
WHERE e.MemberID IN
    (SELECT m.MemberID
     FROM Member m
     WHERE m.MemberType = 'Senior');
```

The inner section in this type of query must return just a single column. IN is expecting a list of single values (in this case, a list of MemberID). If the inner section returns more than one column (for example, SELECT * FROM Member), then we will get an error.

Many nested queries such as this can be written in other ways—often by using an inner join as we discussed earlier in the chapter. Some queries will feel more natural to you one way or the other.

Inner Queries Checking for Existence

Another type of inner query is the one we saw working with the EXISTS keyword. A statement using EXISTS just looks to see whether any rows at all are returned by the inner query. The actual values or numbers of rows returned are not important. The query that follows returns any rows from the Member table where we can find a corresponding row in the Entry table for that member:

```
SELECT m.Lastname, m.FirstName
FROM Member m
WHERE EXISTS
    (SELECT * FROM Entry e
     WHERE e.MemberID = m.MemberID);
```

Because the actual values retrieved by the inner query are not important, the inner query often has the form SELECT * FROM.

Another feature of this type of query is that the inner and outer sections are usually correlated. By this we mean that the WHERE clause in the inner section refers to values in the table in the outer section. In this case the inner query is checking if the current row in the Entry table has the same MemberID as the member currently under consideration in the outer query. I find the easiest way to visualize this is as illustrated in Figure 4-5.

It is difficult to think of a sensible EXISTS query that doesn't correlate values in the inner and outer sections. Consider what the following query will return:

```
SELECT m.Lastname, m.FirstName
FROM Member m
WHERE EXISTS
    (SELECT * FROM Entry e);
```

The query above doesn't really make any sense. It says to write out each member's names if there is a row in the Entry table (any row!). If the Entry table is empty, we will get nothing returned; otherwise, we will get all the names of all the members. I can't think why you'd ever want to do that. EXISTS queries are useful when we are looking for matching values somewhere else, and that is why the SELECT condition needs to compare values from both the inner and outer sections.

It is interesting to compare the following two queries. They both return the names of members who have entered a tournament, but the results are slightly different. The first uses an EXISTS clause:

```
SELECT m.Lastname
FROM Member m
WHERE EXISTS
    (SELECT * FROM Entry e
    WHERE e.MemberID = m.MemberID);
```

The second uses an INNER JOIN:

```
SELECT m.LastName
FROM Member m INNER JOIN Entry e ON  e.MemberID = m.MemberID;
```

The difference between the two queries is the number of rows that are returned.

The first query inspects each row in the Member table just once and returns the last name if there exists at least one entry for that member in the Entry table. The last name for any member will be written out only once.

The second query forms a join between the two tables that will consist of every combination of rows in Member and Entry with the same MemberID. The name for a particular member will be written out as many times as the number of tournaments he or she entered.

It's a subtle difference, but an important one – especially if you are wanting to count the returned rows. Adding DISTINCT in the SELECT clause of the second example will make the results of the two queries the same.

Using Subqueries for Updating

This book is mainly about queries for retrieving data, but many of the same ideas can be used for updating data and adding or deleting records. In Chapter 1 we looked at simple queries such as updating the phone number of a particular member, as shown here:

```
UPDATE Member m
SET m.Phone = '875076'              .
WHERE m.MemberID = 118;
```

We also looked at inserting and deleting rows from a table. To insert a row we list the columns we are providing values for and then the values, as in the following:

```
INSERT INTO Entry (MemberID, TourID, Year)
VALUES (153, 25, 2016);
```

Now, let's consider a situation where we want to add an entry for tournament 25 in 2016 for each of the juniors in the club. We want to add a set of rows to the Entry table, as shown in Figure 4-8, where the left column has the member IDs for each of the juniors and the next two columns are the specific tournament (25) and year (2016) for each entry.

MemberID ▾	TourID ▾	Year ▾
118	25	2016
228	25	2016
286	25	2016
414	25	2016
469	25	2016

Figure 4-8. *Rows to be added to Entry table*

We can write an SQL query to return a set of rows like those in Figure 4-8:

```
SELECT m.MemberID, 25, 2016
FROM Member m
WHERE m.MemberType = 'Junior';
```

This query is a little different from others we have looked at because it has constants in the SELECT clause. It will construct a row for each junior member with the member's ID and the two constants 25 (for the tournament) and 2016 (for the year).

We can now use the preceding query as a subquery in our INSERT query. Rather than provide just one value with the VALUES keyword, we can provide a set of values resulting from the subquery. In the following query, the inner SELECT will produce the set of rows seen in Figure 4-8, and the outer INSERT will put them in the Entry table:

```
INSERT INTO Entry (MemberID, TourID, Year)
    -- create an entry in tournament 25, 2016 for each Junior
    SELECT MemberID, 25, 2016
    FROM Member
    WHERE MemberType = 'Junior';
```

The same potential for using subqueries applies to other updating issues. Say, for the purposes of finding an example, that after entering data in the Entry table for the 2016 social tournament at Kaiapoi (tournament 25) you realize that only players with handicaps of 20 or more were allowed to enter. You could use a subquery to delete entries for members with handicaps less than 20:

```
DELETE FROM Entry
WHERE TourID = 25 AND Year = 2016 AND
MemberID IN
    (SELECT MemberID FROM Member WHERE Handicap < 20);
```

Summary

We can use subqueries along with the keywords IN and EXISTS in many situations. Here is a summary of the situations we have looked at in this chapter.

Examples of Different Types of Subqueries

Many nested queries can be written in alternative ways. In Chapter 9, we will look at performance issues relating to different ways of expressing queries, but in general you should use the way that feels most natural to you when designing a query. Here are some examples of nested queries and alternate ways of expressing them.

A subquery returning a single value

Find the tournaments that member Cooper has entered:

```
SELECT e.TourID, e.Year FROM Entry e WHERE e.MemberID =
    (SELECT m.MemberID FROM Member m
    WHERE m.LastName = 'Cooper');
```

An alternative way to write the preceding query is to use a join:

```
SELECT e.TourID, e.Year
FROM Entry e INNER JOIN Member m ON e.MemberID = m.MemberID
WHERE m.LastName = 'Cooper';
```

A subquery returning a set of single values

Find all the entries for an Open tournament:

```
SELECT *
FROM Entry e
WHERE e.TourID IN
    (SELECT t.TourID FROM Tournament t
    WHERE t.TourType = 'Open');
```

The preceding query can be replaced with:

```
SELECT e.MemberID, e.TourID, e.Year
FROM Entry e INNER JOIN Tournament t ON e.TourID = t.TourID
WHERE t.TourType = 'Open';
```

A subquery checking for existence

Find the names of members that have entered any tournament:

```
SELECT m.LastName, m.FirstName
FROM Member m
WHERE EXISTS
    (SELECT * FROM Entry e
    WHERE e.MemberID = m.MemberID);
```

This can be replaced with:

```
SELECT DISTINCT m.LastName, m.FirstName
FROM Member m INNER JOIN Entry e ON e.MemberID = m.MemberID;
```

Examples of Different Uses for Subqueries

Subqueries can be used in many situations, including the following:

Constructing queries with negatives

Find the names of members who have not entered a tournament:

```
SELECT * FROM Member m
WHERE NOT EXISTS
    (SELECT * FROM Entry e
    WHERE e.MemberID = m.MemberID);
```

Comparing values with the results of aggregates

Find the names of members with handicaps less than the average:

```
SELECT m.LastName, m.FirstName FROM Member m WHERE m.Handicap <
    (SELECT AVG(Handicap) FROM Member);
```

Update data

Add a row in the Entry table for every junior for tournament 25 in 2016:

```
INSERT INTO Entry (MemberID, TourID, Year)
    SELECT MemberID, 25, 2016
    FROM Member WHERE MemberType = 'Junior';
```

CHAPTER 5

■ ■ ■

Self Joins

When we select a subset of rows based on a condition in a WHERE clause, the condition is evaluated for each row independently. An example might be a query to find all the members who have entered tournament 36. The condition TourID = 36 can be evaluated for each row in the Entry table to achieve the required result. However, if we want to find members who have entered both tournaments 36 and 24, we cannot do this by inspecting just one row of the Entry table. We need to find two rows (or entries) for the same member—one for each of the specified tournaments. A simple WHERE clause cannot achieve this.

In this chapter we will look at self joins. With a join between two tables, we first make a Cartesian product that gives us a combination of rows from each table. In a self join, we do the same thing but with two copies of the same table. This provides us with every combination of pairs of rows from the original table. This is one way to write a query that needs information from more than one row in a table to satisfy some condition. It will enable us to answer questions involving the word *both*; for example, "Which members entered *both* these tournaments?" Self joins will also allow us to carry out queries on tables involved in self relationships. We'll look at self relationships first.

Self Relationships

Let's add some more information to our Member table. Suppose some members have coaches assigned to them. How do we represent that in the class diagrams we talked about in Chapter 1? We could take the approach shown in Figure 5-1, with two classes: Member and Coach. Recall what the lines and numbers mean. From left to right, a coach might have several members to train (the 0..n nearest the Member class). From right to left, a particular member might have a single coach or no coach (the 0..1 nearest the Coach class).

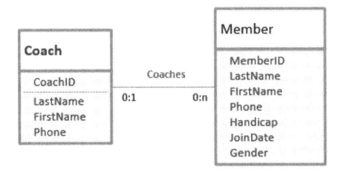

Figure 5-1. *Data model for coaches and members (not recommended!)*

© Clare Churcher 2016

C. Churcher, *Beginning SQL Queries*, DOI 10.1007/978-1-4842-1955-3_5

The problem with the model in Figure 5-1 is that coaches, in all probability, are members of the club. When we implement this model with a Coach table and a Member table, some people will have a row recording their details in each table. For example, Brenda Nolan has a row in the Member table. When she takes up a role as coach, we also would need a row about her in the Coach table. Now if Brenda gets a new phone number, someone has to remember to change it in both tables. In all likelihood this won't happen, and we will end up with the old number in one of the tables.

In this example we don't actually have two separate classes of members and coaches. We have just one class of members, some of whom coach other members. This self relationship is shown in Figure 5-2.

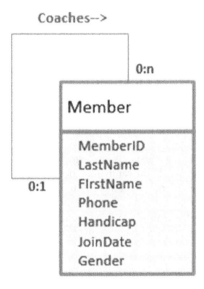

Figure 5-2. *Data model for members coaching other members*

The relationship line in Figure 5-2 can be read in a clockwise direction to say that a particular member might coach several other members or none (0..n). In the other direction, we can read that a particular member might have one coach or none (0..1).

In Chapter 1 we showed how to represent a 1–Many relationship by adding a column to the table at the 1 end of the relationship, which will have values from the primary key of the table at the other end. The model in Figure 5-2 is exactly the same type of 1–Many relationship, except that we have the same table at each end, hence a *self relationship*. To represent the relationship we can add a column, Coach, in the Member table, as shown in Figure 5-3. The values in the Coach field must also exist in the key field MemberID.

MemberID ▾	LastName ▾	FirstName ▾	Handicap ▾	MemberType ▾	Gender ▾	Coach ▾
118	McKenzie	Melissa	30	Junior	F	153
138	Stone	Michael	30	Senior	M	
153	Nolan	Brenda	11	Senior	F	
176	Branch	Helen		Social	F	
178	Beck	Sarah		Social	F	
228	Burton	Sandra	26	Junior	F	153
235	Cooper	William	14	Senior	M	153
239	Spence	Thomas	10	Senior	M	
258	Olson	Barbara	16	Senior	F	
286	Pollard	Robert	19	Junior	M	235
290	Sexton	Thomas	26	Senior	M	235
323	Wilcox	Daniel	3	Senior	M	
331	Schmidt	Thomas	25	Senior	M	153
332	Bridges	Deborah	12	Senior	F	235
339	Young	Betty	21	Senior	F	
414	Gilmore	Jane	5	Junior	F	153
415	Taylor	William	7	Senior	M	235
461	Reed	Robert	3	Senior	M	235
469	Willis	Carolyn	29	Junior	F	
487	Kent	Susan		Social	F	

Figure 5-3. Column Coach added to the Member table

The first row in the table in Figure 5-3 tells us that Melissa is coached by member 153, and we can see from the third line of the table that member 153 is Brenda. We need the value in the Coach field to be constrained to being one of our existing members so that we cannot accidentally add an invalid member number in the Coach column. We can do this by making the Coach field a foreign key. Recall from Chapter 1 that a foreign key is a field where any non-empty values in the field must already exist as a primary key in another table. For the table in Figure 5-3, MemberType is a foreign key referring to the Type table, meaning that any value in the MemberType column must already exist in the Type table. For the Coach column, the "other" table is the Member table itself. The following SQL statement shows how we would use the ALTER command to add the new foreign key column Coach:

```
ALTER TABLE Member
ADD Coach INT FOREIGN KEY REFERENCES Member;
```

With the modified Member table, we now can answer several different types of questions. For example:

- What are the names of the coaches?

- What is the name of Jane Gilmore's coach?

- Is anyone being coached by someone with a higher handicap?

- Are any women being coached by men?

None of these questions can be answered by inspecting a single row in the table. What we require is a *self join* on the Member table. The easiest way to understand a self join is to see how we make one.

Creating a Self Join

Recall from Chapter 3 the definition of a join between two tables: a Cartesian product (every combination of rows from each table) followed by selecting a subset of those rows that satisfy some join condition. For a self join, we think of two copies of the same table. In Figure 5-4, we see part of the Cartesian product between two copies of the Member table. To distinguish the different elements of the product, I've given the first copy an alias, m, and the second a different alias, c (you'll see why in a minute). In the small section of the Cartesian product visible in Figure 5-4, we see the first row (Melissa) from copy m paired with each of the rows from copy c. Some of the headings of the columns are truncated, as it was getting rather wide.

Rows from first copy of Member (m)							Rows from second copy of Member (c)						
m.M ▾	m.LastN; ▾	m.FirstN ▾	m.Men ▾	m.F ▾	m.G ▾	m.C ▾	c.M ▾	c.LastNar ▾	c.FirstN ▾	c.Mem ▾	c.F ▾	c. ▾	c.C ▾
118	McKenzie	Melissa	Junior	30	F	153	118	McKenzie	Melissa	Junior	30	F	153
118	McKenzie	Melissa	Junior	30	F	153	138	Stone	Michael	Senior	30	M	
118	McKenzie	Melissa	Junior	30	F	153	153	Nolan	Brenda	Senior	11	F	
118	McKenzie	Melissa	Junior	30	F	153	176	Branch	Helen	Social		F	
118	McKenzie	Melissa	Junior	30	F	153	178	Beck	Sarah	Social		F	
118	McKenzie	Melissa	Junior	30	F	153	228	Burton	Sandra	Junior	26	F	153
118	McKenzie	Melissa	Junior	30	F	153	235	Cooper	William	Senior	14	M	153
118	McKenzie	Melissa	Junior	30	F	153	239	Spence	Thomas	Senior	10	M	
118	McKenzie	Melissa	Junior	30	F	153	258	Olson	Barbara	Senior	16	F	
118	McKenzie	Melissa	Junior	30	F	153	286	Pollard	Robert	Junior	19	M	235
118	McKenzie	Melissa	Junior	30	F	153	290	Sexton	Thomas	Senior	26	M	235
118	McKenzie	Melissa	Junior	30	F	153	323	Wilcox	Daniel	Senior	3	M	
118	McKenzie	Melissa	Junior	30	F	153	331	Schmidt	Thomas	Senior	25	M	153
118	McKenzie	Melissa	Junior	30	F	153	332	Bridges	Deborah	Senior	12	F	235
118	McKenzie	Melissa	Junior	30	F	153	339	Young	Betty	Senior	21	F	
118	McKenzie	Melissa	Junior	30	F	153	414	Gilmore	Jane	Junior	5	F	153
118	McKenzie	Melissa	Junior	30	F	153	415	Taylor	William	Senior	7	M	235
118	McKenzie	Melissa	Junior	30	F	153	461	Reed	Robert	Senior	3	M	235
118	McKenzie	Melissa	Junior	30	F	153	469	Willis	Carolyn	Junior	29	F	
118	McKenzie	Melissa	Junior	30	F	153	487	Kent	Susan	Social		F	

m.Coach   c.MemberID

Figure 5-4. *Cartesian product between two copies of the Member table*

For queries about coaching, the interesting rows from the Cartesian product are those where the value of Coach from m is the same as MemberID from c. In Figure 5-4, you can see that the third line contains information about Melissa (from the m copy of Member) and information about her coach (from the c copy of Member). Now the choice of aliases becomes clear: m for columns about a member; c for the columns about that member's coach. Choosing helpful aliases can make understanding self joins much easier. The rows we would like to select from the Cartesian product are those satisfying m.Coach = c.MemberID. This is the join condition required to find information about members and their coaches. The SQL for the self join is:

```
SELECT *
FROM Member m INNER JOIN Member c ON m.Coach = c.MemberID;
```

The table resulting from the self join is shown in Figure 5-5.

	Information about a member (m)							Information about their coach (c)					
m.M ◄	m.LastN₂ ▾	m.FirstN ▾	m.Men ▾	m.F ▾	m.G ▾	m.C ▾	c.M ◄	c.LastNar ▾	c.FirstN ▾	c.Mem ▾	c.F ▾	c. ▾	c.C ▾
118	McKenzie	Melissa	Junior	30	F	153	153	Nolan	Brenda	Senior	11	F	
228	Burton	Sandra	Junior	26	F	153	153	Nolan	Brenda	Senior	11	F	
235	Cooper	William	Senior	14	M	153	153	Nolan	Brenda	Senior	11	F	
286	Pollard	Robert	Junior	19	M	235	235	Cooper	William	Senior	14	M	153
290	Sexton	Thomas	Senior	26	M	235	235	Cooper	William	Senior	14	M	153
331	Schmidt	Thomas	Senior	25	M	153	153	Nolan	Brenda	Senior	11	F	
332	Bridges	Deborah	Senior	12	F	235	235	Cooper	William	Senior	14	M	153
414	Gilmore	Jane	Junior	5	F	153	153	Nolan	Brenda	Senior	11	F	
415	Taylor	William	Senior	7	M	235	235	Cooper	William	Senior	14	M	153
461	Reed	Robert	Senior	3	M	235	235	Cooper	William	Senior	14	M	153

m.Coach = c.MemberID

Figure 5-5. *Self join on Member table to retrieve information about members and their coaches*

Now that we have the results of the self join, we can answer the questions posed in the previous section about coaching. The trickiest part of all this was recognizing that maintaining information about members and coaches is a self relationship and designing the Member table appropriately in the first place.

Queries Involving a Self Join

With the joined table in Figure 5-5 as our base, we can answer all sorts of questions by simply selecting subsets of rows and projecting the appropriate columns. Whenever I need to do queries involving self joins, I usually perform the join first, retaining all the rows and columns as in Figure 5-5. With the joined table (or a quick sketch of the columns) in front of me, the way forward is usually relatively simple. Let's see how this works with a few questions.

What Are the Names of the Coaches?

Looking at Figure 5-5, we can see that the names of the coaches are in the columns coming from the c part of the join. We just want a list of the names in the columns c.LastName and c.FirstName so those columns can be included in the SELECT clause. We don't want the names repeated, so we use the keyword DISTINCT. The following SQL statement will return the names of the two coaches, Brenda Nolan and William Cooper.

```
SELECT DISTINCT c.FirstName, c.LastName
FROM Member m INNER JOIN Member c ON m.Coach = c.MemberID;
```

Who Is Being Coached by Someone with a Higher Handicap?

To find out who is being coached by someone with a higher handicap, we need to compare the handicap of the member (m.Handicap) with the handicap of that member's coach (c.Handicap). What is required is a WHERE clause after the join clause to find where the member's handicap is less than the coach's handicap:

```
SELECT *
FROM Member m INNER JOIN Member c ON m.Coach = c.MemberID
WHERE m.Handicap < c.Handicap;
```

For the data in Figure 5-5, this will retrieve the data in the last four rows. (You don't have to be a great golfer to be a good coach!) Having done the join and selected the appropriate rows, we can then choose which columns we want to appear in the final result and list them in the SELECT clause.

List the Names of All Members and the Names of Their Coaches

Listing the names of members and their coaches sounds pretty trivial, but if we are not careful, we can get it wrong. A first thought might be to project just the four columns containing the names of member and coach from the joined table in Figure 5-5. However, there are only 10 rows in the joined table, whereas there are 20 members in the Member table. The issue here is that not all the members have coaches. We looked at situations like this in the section on outer joins in Chapter 3.

To recap, let's go back to the Cartesian product of two copies of the Member table, but look at some rows involving a member with no coach, as shown in Figure 5-6.

Information about a member (m)   Information about their coach (c)

m.M ▾	m.LastN ▾	m.FirstN ▾	m.Men ▾	m.F ▾	m.G ▾	m.Coach ▾	c.MemberID ▾	c.LastNar ▾	c.Firstℕ ▾	c.Mem ▾	c.I ▾	c. ▾	c.C ▾
138	Stone	Michael	Senior	30	M		118	McKenzie	Melissa	Junior	30	F	153
138	Stone	Michael	Senior	30	M		138	Stone	Michael	Senior	30	M	
138	Stone	Michael	Senior	30	M		153	Nolan	Brenda	Senior	11	F	
138	Stone	Michael	Senior	30	M		176	Branch	Helen	Social		F	
138	Stone	Michael	Senior	30	M		178	Beck	Sarah	Social		F	
138	Stone	Michael	Senior	30	M		228	Burton	Sandra	Junior	26	F	153
138	Stone	Michael	Senior	30	M		235	Cooper	William	Senior	14	M	153

m.Coach = c.MemberID
Never satisfied when m.Coach is null

Figure 5-6. *Part of the Cartesian product between two copies of the Member table*

The join condition (m.Coach = c.MemberID) is never satisfied for a member with a null in the Coach column, so all those members will be missing from our joined table. We just need to be careful to understand what we really want. Do we want a list of all the members with coaches (10 rows), or a list of all the members along with their coach's name if they have one (20 rows)? If it's the latter, we need an outer join. We need to see the name of each member (from the m copy of the Member table), along with the name of his coach, if any (from the c copy). The SQL for this outer join is:

```
SELECT m.LastName AS MemberLast, m.FirstName AS MemberFirst,
       c.LastName AS CoachLast, c.FirstName AS CoachFirst
FROM Member m LEFT OUTER JOIN Member c ON m.Coach = c.MemberID;
```

In the preceding query we have given each output attribute a *column alias*. A column alias temporarily renames a column in order to improve the readability of the output. In this case it helps the reader distinguish which name belongs to whom, as shown in Figure 5-7. Without the aliases, the attributes would be labelled as m.LastName and c.LastName and so on, which are not quite so easy to understand. Recall from Chapter 3 that for a left outer join, where there is no matching row from the right-hand table, those columns will be filled with nulls. Figure 5-7 shows the output of the left outer join.

MemberLast ▾	MemberFirst ▾	CoachLast ▾	CoachFirst ▾
McKenzie	Melissa	Nolan	Brenda
Stone	Michael		
Nolan	Brenda		
Branch	Helen		
Beck	Sarah		
Burton	Sandra	Nolan	Brenda
Cooper	William	Nolan	Brenda
Spence	Thomas		
Olson	Barbara		
Pollard	Robert	Cooper	William
Sexton	Thomas	Cooper	William
Wilcox	Daniel		
Schmidt	Thomas	Nolan	Brenda
Bridges	Deborah	Cooper	William
Young	Betty		
Gilmore	Jane	Nolan	Brenda
Taylor	William	Cooper	William
Reed	Robert	Cooper	William
Willis	Carolyn		
Kent	Susan		

Figure 5-7. *Left outer join to list all members and coaches*

Who Coaches the Coaches, or Who Is My Grandmother?

The self join between two copies of the Member table shows us one level of members and coaches. If we look at the rows in Figure 5-7, we can see that Thomas Sexton is coached by William Cooper, who is in turn coached by Brenda Nolan, who doesn't have a coach. The hierarchy isn't all that interesting for this problem, but there are several analogous situations where the hierarchy is of considerable interest. Genealogy is one. Consider the data model and part of the Person table in Figure 5-8. For the sake of keeping things really simple, we will consider only a tiny bit of information about just women and birth mothers.

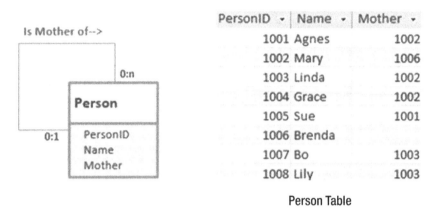

PersonID ▾	Name ▾	Mother ▾
1001	Agnes	1002
1002	Mary	1006
1003	Linda	1002
1004	Grace	1002
1005	Sue	1001
1006	Brenda	
1007	Bo	1003
1008	Lily	1003

Person Table

Figure 5-8. Data model for women and their birth mothers

The relationship in Figure 5-8 can be read clockwise as "a person can be the mother of several other people" and in the other direction as "a person has at most one mother and might have none." Now. in real life, that last statement doesn't sound right—surely everyone has a mother. However, as with all databases, this database is only an approximation of the complexities of real life, and it can only keep data that is available. Unless we trace everyone back to the primeval slime, there will be some people in our table whose mother we do not know. Brenda is one. The table and model in Figure 5-8 have exactly the same structure as our member and coach example, but a question like "Who is Sue's grandmother?" seems a bit more likely than "Who coaches my coach?"

So, how do we get information about people along with information about their mothers? Just as in the previous section, we need to join the Person table to itself. (Don't forget to make the join an outer join so you don't lose Brenda.) The SQL is:

```
SELECT *
FROM Person p LEFT OUTER JOIN Person m on p.Mother = m.ID;
```

The Access diagrammatic interface for the join is shown in Figure 5-9, along with the resulting table. I've given the first copy of the table the alias p for *person* and the second copy the alias m for *mother*.

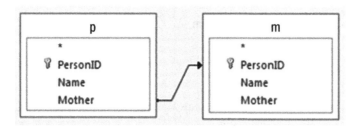

p.PersonID ▾	p.Name ▾	p.Mother ▾	m.PersonID ▾	m.Name ▾	m.Mother ▾
1001	Agnes	1002	1002	Mary	1006
1002	Mary	1006	1006	Brenda	
1003	Linda	1002	1002	Mary	1006
1004	Grace	1002	1002	Mary	1006
1005	Sue	1001	1001	Agnes	1002
1006	Brenda				
1007	Bo	1003	1003	Linda	1002
1008	Lily	1003	1003	Linda	1002

Figure 5-9. *Finding people and their mothers: Access diagram for the left outer join and the resulting table*

Now, what about going back to the previous generation? For that we need to perform another left outer join between the result table in Figure 5-9 and another copy of the People (with the alias g for *grandmother*). The SQL for the two left outer joins is:

```
SELECT *
FROM (Person p LEFT JOIN Person m ON p.Mother = m.ID)
     LEFT JOIN Person g ON m.Mother = g.ID;
```

The resulting table is shown in Figure 5-10.

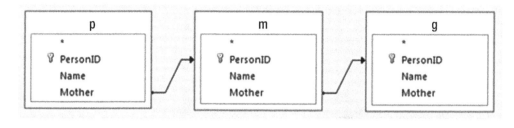

p.PersonID ▾	p.Name ▾	p.Mother ▾	m.PersonID ▾	m.Name ▾	m.Mother ▾	g.PersonID ▾	g.Name ▾	g.Mother ▾
1001	Agnes	1002	1002	Mary	1006	1006	Brenda	
1002	Mary	1006	1006	Brenda				
1003	Linda	1002	1002	Mary	1006	1006	Brenda	
1004	Grace	1002	1002	Mary	1006	1006	Brenda	
1005	Sue	1001	1001	Agnes	1002	1002	Mary	1006
1006	Brenda							
1007	Bo	1003	1003	Linda	1002	1002	Mary	1006
1008	Lily	1003	1003	Linda	1002	1002	Mary	1006

Figure 5-10. *Finding three generations: Access diagram for the left outer joins and the resulting table*

Clearly, we can keep making more and more self joins until we run out of generations. These sorts of hierarchical queries are likely to turn up whenever we have self relationships. One small catch is that we must specify the number of joins in each query. Standard SQL doesn't have the notion of a query that automatically keeps doing the self joins until it runs out of generations, such as "Find all my female ancestors"; however, some implementations do support this.[1]

An Outcome Approach to Self Joins

The questions in the previous sections were all quite easy to answer once we realized we needed self joins. This was an example of the process approach – what operations do we need to perform? Sometimes, however, these realizations don't always come when you need them. Whenever my mind goes blank when faced with a query, I resort to an outcome approach.

Let's look at our Member table again and ask a simple question: Who is Melissa's coach? Don't think about relationships or joins, just look at the data from a layman's perspective. In Figure 5-11, you can see how to figure out the answer, even if you have never heard of a self join (most people haven't).

---

[1]Some implementations of SQL do support recursive queries that can track through self relationships. Check your documentation for key phrases like WITH or CONNECT BY.

MemberID ▾	LastName ▾	FirstName ▾	Handicap ▾	Coach ▾
m ☞ 118 McKenzie		Melissa	30	153
138 Stone		Michael	30	
c ☞ 153 Nolan		Brenda	11	
176 Branch		Helen		
178 Beck		Sarah		
228 Burton		Sandra	26	153
235 Cooper		William	14	153
239 Spence		Thomas	10	
258 Olson		Barbara	16	
286 Pollard		Robert	19	235

Figure 5-11. *Finding Melissa's coach*

To find Melissa's coach, we first find the row for Melissa (m in Figure 5-11) and then note that her coach is member 153. Then we find another row (c for coach) that has the MemberID value of 153; we can see that Melissa's coach is Brenda. You don't need to know anything about self relationships or foreign keys or joins to figure that out. But once you have that logic clearly in your mind, you can write it down in natural language, and then the translation to SQL is pretty straightforward.

Let's write a description of Figure 5-11:

> *I need to look at two rows (m and c) in the Member table, and I want to write out c.FirstName where c.MemberID has the same value as m.Coach and m.FirstName is 'Melissa'*

And here is the corresponding SQL:

```
SELECT c.FirstName
FROM Member m, Member c
WHERE c.MemberID = m.Coach AND m.FirstName = 'Melissa';
```

So, how does this output approach correspond to the process approach we considered earlier? As you might expect, the preceding SQL is just an alternative way of stating the same query as the one where we used the self join. In the preceding SQL statement, the middle line is the Cartesian product between two copies of the Member table, and the first part of the WHERE clause is the join condition. The statement FROM Member m, Member c WHERE c.MemberID = m.Coach is just another way of expressing the self join we used in the previous sections.

Let's try one of the other queries using an outcome approach: Who is being coached by someone with a higher handicap? The picture I would need in my head to answer this question is shown in Figure 5-12.

MemberID ▾	LastName ▾	FirstName ▾	Handicap ▾	Coach ▾
118	McKenzie	Melissa	30	153
138	Stone	Michael	30	
153	Nolan	Brenda	11	
176	Branch	Helen		
178	Beck	Sarah		
228	Burton	Sandra	26	153
(235)	Cooper	William	14	153
239	Spence	Thomas	10	
258	Olson	Barbara	16	
286	Pollard	Robert	19	235
290	Sexton	Thomas	26	235
323	Wilcox	Daniel	3	
331	Schmidt	Thomas	25	153
332	Bridges	Deborah	12	(235)
339	Young	Betty	21	
414	Gilmore	Jane	5	153
415	Taylor	William	7	235
461	Reed	Robert	3	235
469	Willis	Carolyn	29	
487	Kent	Susan		

c ☞ (row 235)   m ☞ (row 332)   m.Handicap < c.Handicap

Figure 5-12. *Finding members who are coached by someone with a higher handicap*

We can see that Deborah, whose handicap is 12, is being coached by member 235. Member 235, William, has a handicap of 14, so Deborah satisfies our criteria. Here is the more general statement representing the logic depicted in Figure 5-12:

I'm going to look at every row (m) in the Member table and will write out m.FirstName and m.LastName if there exists some other row (c) in the Member table where c.MemberID is the same as m.Coach and m.Handicap is less than c.Handicap

The SQL follows in a straightforward manner:

```
SELECT m.FirstName, m.LastName
FROM Member m, Member c
WHERE c.MemberID = m.Coach AND m.Handicap < c.Handicap;
```

Once again, you can see the equivalent of the self join in the preceding query (FROM Member m, Member c WHERE c.MemberID = m.Coach). The usefulness of this outcome approach is that you don't need to understand what a self join is, nor must you make the mental leap that you need one. By thinking in terms of virtual fingers and which rows are involved in helping you with your decision, you can sketch a statement of the criteria. The SQL usually follows quite easily from that.

Questions Involving "Both"

In the "Avoiding Common Mistakes" section of Chapter 2, we looked at a questions such as, "Which members have entered *both* tournaments 24 and 36?" To recap, I've reproduced the Entry table in Figure 5-13.

MemberID ▾	TourID ▾	Year ▾
118	24	2014
228	24	2015
228	25	2015
228	36	2015
235	38	2013
235	38	2015
235	40	2014
235	40	2015
239	25	2015
239	40	2013
258	24	2014
258	38	2014
286	24	2013
286	24	2014
286	24	2015
415	24	2015
415	25	2013
415	36	2014
415	36	2015
415	38	2013
415	38	2015
415	40	2013
415	40	2014
415	40	2015

Figure 5-13. Entry table

A common first attempt at an SQL statement to find entries in both tournaments is the following:

```
-- Will not produce the desired result
SELECT e.MemberID
FROM Entry e
WHERE e.TourID = 24 AND e.TourID = 36;
```

Remember that a WHERE condition is applied to each row of the table individually. The condition (e.TourID = 24 AND e.TourID = 36) is never true for any individual row, as each row has only a single value for TourID. The preceding query will never return any rows because the value in TourID cannot be two different things (24 and 36) simultaneously. Such a query can be quite dangerous, because the user may interpret the empty result as meaning that no members have entered both tournaments, whereas the query statement is actually incorrect.

To answer the question, we need to look at more than one row in the Entry table. I find an outcome approach to be the most natural for dealing with questions involving "both."

An Outcome Approach to Questions Involving "Both"

The picture I need in my head to answer "Which members have entered both tournaments 24 and 36?" is shown in Figure 5-14.

MemberID ▾	TourID ▾	Year ▾
118	24	2014
e1 ☞ (228)	[24]	2015
228	25	2015
e2 ☞ (228)	[36]	2015
235	38	2013
235	38	2015
235	40	2014
235	40	2015

Figure 5-14. *Which members have entered both tournaments 24 and 36?*

Looking at Figure 5-14, it is pretty clear that member 228 has entered both the tournaments. We are to looking for two rows (two fingers, e1 and e2) with matching MemberID values and where the rows have the required two TourID values.

A more general expression of the logic displayed in Figure 5-14 is:

I'm going to look at every row (e1) in the Entry table. I'll write out that row's member ID if TourID has the value 24 and I can also find another row (e2) in the Entry table with the same value for MemberID and that has 36 as the value for TourID.

The SQL follows from here. If you have trouble with it, refer to Figure 5-14.

```
SELECT e1.MemberID
FROM Entry e1, Entry e2
WHERE e1.MemberID = e2.MemberID
    AND e1.TourID = 24 AND e2.TourID = 36;
```

A Process Approach to Questions Involving "Both"

As always, we have several ways to think about a query. Take a look at the middle two lines of the last query. FROM Entry e1, Entry e2 is a Cartesian product (which will give us every combination of pairs of rows), followed by selecting a subset of rows satisfying (WHERE e1.MemberID = e2.MemberID). This is a join. In fact, it is a self join between two copies of the Entry table. Part of the join between two copies of the Entry table is shown in Figure 5-15.

From copy e1 of Entry			From copy e2 of Entry		
e1.MemberID ▾	e1.TourID ▾	e1.Year ▾	e2.MemberID ▾	e2.TourID ▾	e2.Year ▾
118	24	2014	118	24	2014
228	24	2015	228	24	2015
228	25	2015	228	24	2015
228	36	2015	228	24	2015
228	24	2015	228	25	2015
228	25	2015	228	25	2015
228	36	2015	228	25	2015
228	24	2015	228	36	2015
228	25	2015	228	36	2015
228	36	2015	228	36	2015
235	38	2013	235	38	2013
235	38	2015	235	38	2013
235	40	2014	235	38	2013
235	40	2015	235	38	2013
235	38	2013	235	38	2015

Join condition
e1.MemberID = e2.MemberID

Figure 5-15. Part of the self join between two copies of the Entry table

The self join in Figure 5-15 shows those combinations of rows from the Entry table for the same member. For example, we can see every combination of rows involving member 228. We can use this self join to answer the question about members who have entered both tournaments 24 and 36. We just need to find a row that has 24 from the first copy and 36 from the second copy (or vice versa) – that is, e1.TourID = 24 AND e2.TourID = 36.

The SQL for this self join followed by the WHERE clause to select the rows with the appropriate values of TourID is shown here:

```
SELECT e1.MemberID
FROM Entry e1 INNER JOIN Entry e2 ON e1.MemberID = e2.MemberID
WHERE e1.TourID = 24 AND e2.TourID = 36;
```

If you compare the two queries for finding the entries in both tournaments 24 and 26, you will see how similar they are. They will both produce exactly the same result. You will probably find one or the other to be more intuitive.

Summary

Many queries require us to obtain information from two rows of a table. This turns up in a number of situations. The main ones are where we have self relationships or where there are questions involving the word "both." We have looked at both process approaches and outcome approaches to these queries. Both resulted in very similar-looking SQL statements that return the same output. Having the two different approaches is helpful for those occasions when the query statement is not immediately obvious.

Self Relationships

We have a self relationship when different instances of a class are related to each other. In the example in this chapter, we had that some members are coaches of other members.

From a process perspective, queries about coaches or coaching relationships require self joins, which take two copies of the table and join them. In the following example, the copy of the Member table with the information about the member has the alias m, and the copy with information about the coach has the alias c:

```
SELECT m.LastName, m.FirstName, c.LastName, c.FirstName
FROM Member m INNER JOIN Member c ON m.Coach = c.MemberID
```

Alternatively, from an output approach we might come up with this equivalent query:

```
SELECT m.FirstName, m.LastName, c.LastName, c.FirstName
FROM Member m, Member c
WHERE c.MemberID = m.Coach
```

Both these queries can form the basis of queries to answer a number of questions about coaching.

Questions Involving the Word "Both"

Questions with the word "both" often mean we need to look at two rows in a table. In our example, we wanted to find the MemberID of members who have entered both tournaments 24 and 36.

From an outcome approach we needed to find two rows in the Entry table (e1 and e2) for the same member. One of the rows needed to be for tournament 24 and the other for tournament 36. The following shows the outcome-based SQL query:

```
SELECT e1.MemberID
FROM Entry e1, Entry e2
WHERE e1.MemberID = e2.MemberID AND e1.TourID = 24 AND e2.TourID = 36;
```

Alternatively, from a process approach we might recognize the need for a self join between two copies of the Entry table, which is done using the join condition e1.MemberID = e2.MemberID. This would need to be followed by a WHERE clause to return the rows with the appropriate TourID values.

The self join query equivalent to the preceding query is:

```
SELECT e1.MemberID
FROM Entry e1 INNER JOIN Entry e2 ON e1.MemberID = e2.MemberID
WHERE e1.TourID = 24 AND e2.TourID = 36;
```

CHAPTER 6

■ ■ ■

Multiple Relationships Between Tables

We have looked at simple 1–Many relationships between tables (e.g., each member is associated with one member type), and we have also looked at self relationships (e.g., members may coach other members). Another situation that occurs frequently is where there is more than one relationship between the same two tables.

Two Relationships Between the Same Tables

Let's consider how we might introduce the idea of teams into the golf club database. We can start off by thinking about what basic information we need to keep about a team. Figure 6-1 shows a class representing a simple team along with some rows in a table that represents the class.

Team
TeamName
PracticeNight

TeamName ▾	PracticeNight ▾
TeamA	Tuesday
TeamB	Monday

Figure 6-1. *The Team class and some rows in a Team table*

Now we need to think about relationships between the new Team class and our other classes. The most obvious relationship is that members will play for teams. Figure 6-2 shows a possible class diagram representing this situation.

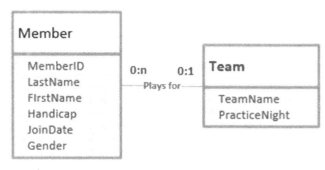

Figure 6-2. *A member can belong to one team*

© Clare Churcher 2016
C. Churcher, *Beginning SQL Queries*, DOI 10.1007/978-1-4842-1955-3_6

Interpreting the class diagram in Figure 6-2 from left to right, we have that a particular member might play on one team (the 1 nearest the Team class), but a member does not need to play for any teams at all (the 0 nearest the Team class). Reading from right to left, we have that a team could have many members playing for it (the n nearest the Member class) but might not have any (the 0 nearest the Member class). That last statement might seem a bit odd, but when we add new teams, or want to start afresh in a new season, a team might not have any members straight away.

To represent a 1–Many relationship, recall from Chapter 1 that we take the primary key from the table at the 1 end of the relationship and add it as a foreign key to the table at the Many end. Figure 6-3 shows a new foreign key field, Team, which refers to the Team table.

MemberID ▾	LastName ▾	FirstName ▾	Team ▾
118	McKenzie	Melissa	
138	Stone	Michael	
153	Nolan	Brenda	TeamB
176	Branch	Helen	
178	Beck	Sarah	
228	Burton	Sandra	
235	Cooper	William	TeamB
239	Spence	Thomas	
258	Olson	Barbara	
286	Pollard	Robert	TeamB
290	Sexton	Thomas	
323	Wilcox	Daniel	TeamA
331	Schmidt	Thomas	
332	Bridges	Deborah	
339	Young	Betty	TeamB
414	Gilmore	Jane	TeamA
415	Taylor	William	TeamA
461	Reed	Robert	TeamA
469	Willis	Carolyn	
487	Kent	Susan	

Figure 6-3. *Foreign key field Team in the Member table*

Another relationship that is likely to occur between Member and Team is that a member might manage a team. Figure 6-4 shows this additional relationship in the class diagram.

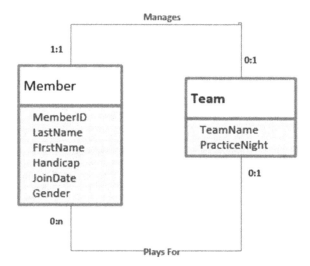

Figure 6-4. *Two relationships between the Member and Team classes*

The top line in Figure 6-4 can be interpreted, from left to right, as stating that a particular member might manage (at most) one team; and from right to left, as that each team has exactly one manager.

This new relationship is a 1–1 relationship. For 1–Many relationships we have always taken the primary key from the one end of the relationship and put it in the table at the other end. This time both ends have a cardinality of 1. We could put a Team_I_Manage column in the Member table or a Manager column in the Team table. The latter is more sensible, as the compulsory Manager attribute is a more important piece of information about teams than the optional Team_I_Manage is for members. Generally, in a 1–1 relationship we take the primary key from the compulsory end (1:1 on the diagram in Figure 6-4) and put that as a foreign key in the other end.

The Team table, with its new Manager foreign key column, is shown along with the Member table in Figure 6-5.

MemberID ▾	LastName ▾	FirstName ▾	Team ▾
118	McKenzie	Melissa	
138	Stone	Michael	
153	Nolan	Brenda	TeamB
176	Branch	Helen	
178	Beck	Sarah	
228	Burton	Sandra	
235	Cooper	William	TeamB
239	Spence	Thomas	
258	Olson	Barbara	
286	Pollard	Robert	TeamB
290	Sexton	Thomas	
323	Wilcox	Daniel	TeamA
331	Schmidt	Thomas	
332	Bridges	Deborah	
339	Young	Betty	TeamB
414	Gilmore	Jane	TeamA
415	Taylor	William	TeamA
461	Reed	Robert	TeamA
469	Willis	Carolyn	
487	Kent	Susan	

Member

TeamName ▾	PracticeNight ▾	Manager ▾
TeamA	Tuesday	239
TeamB	Monday	153

Team

Figure 6-5. *Foreign keys Team in Member table and Manager in Team table to represent the relationships in Figure 6-4*

From the Member table, we can see that four people play for TeamB (Brenda Nolan, William Cooper, Robert Pollard, and Betty Young), and from the Team table, we can see that member 153 (Brenda Nolan) is the manager of TeamB. You will notice that there is nothing in the data model that says whether or not a manager must be a member of the team. TeamB's manager is a member of TeamB, whereas TeamA's manager, 239 (Thomas Spence), is not a member of TeamA. The only constraints implied by the foreign keys are that the manager of a team must be in the Member table and a member can belong only to a team that exists in the Team table.

Some of you may have also realized that making Manager a foreign key does not prevent the same person from managing more than one team. The foreign key constraint does not prevent us from putting member 239 as the manager for both TeamA and TeamB. We have effectively set up a 1-Many relationship between Team and Member for the Manages relationship. If you want to prevent a single member from managing more than one team, you can put a UNIQUE constraint on the Manager column of the Team table. This type of situation is discussed in more depth in my database design book.[1] The following SQL would create a Team table where Manager is a foreign key referring to the Member table and a particular member can only appear once in the Manger column in the table:

```
CREATE TABLE Team (
TeamName CHAR(10) PRIMARY KEY,
PracticeNight CHAR(20),
Manager INT FOREIGN KEY REFERENCES Member UNIQUE);
```

---

[1]Clare Churcher, *Beginning Database Design: From Novice to Professional* (New York: Apress, 2012).

Extracting Information from Multiple Relationships

Now that we have the Team and Member tables and their two relationships (Plays for and Manages), we can start extracting information. If we just consider one relationship at a time, it is relatively straightforward to construct queries. If we want a list of the members who play for a team along with the basic information about their teams from the Team table, we can simply join the Member and Team tables on Team = TeamName as in the SQL query here:

```
SELECT m.MemberID, m.LastName, m.FirstName, m.Team,
       t.TeamName, t.PracticeNight, t.Manager
FROM Member m INNER JOIN Team t ON m.Team = t.TeamName;
```

A graphical representation and the output of the preceding query is shown in Figure 6-6.

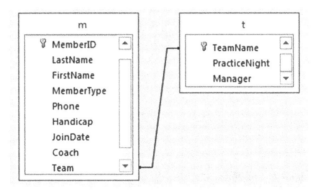

MemberID ▾	LastName ▾	FirstName ▾	Team ▾	TeamName ▾	PracticeNight ▾	Manager ▾
323	Wilcox	Daniel	TeamA	TeamA	Tuesday	239
414	Gilmore	Jane	TeamA	TeamA	Tuesday	239
415	Taylor	William	TeamA	TeamA	Tuesday	239
461	Reed	Robert	TeamA	TeamA	Tuesday	239
153	Nolan	Brenda	TeamB	TeamB	Monday	153
235	Cooper	William	TeamB	TeamB	Monday	153
286	Pollard	Robert	TeamB	TeamB	Monday	153
339	Young	Betty	TeamB	TeamB	Monday	153

Join Condition
m.Team = t.TeamName

Figure 6-6. *Joining Member and Team to get additional information about a member's team*

Similarly, if we want to retrieve information about teams, including the name of the manager, we can join Member and Team on Manager = MemberID:

```
SELECT t.TeamName, t.PracticeNight, t.Manager,
       m.MemberID, m.LastName, m.FirstName
FROM Team t INNER JOIN Member m ON t.Manager = m.MemberID;
```

A graphical representation and the output of the preceding query is shown in Figure 6-7.

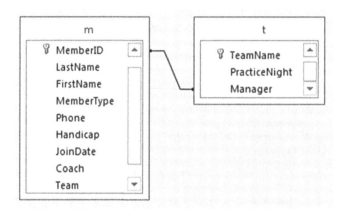

TeamName ▾	PracticeNight ▾	Manager ▾	MemberID ▾	LastName ▾	FirstName ▾
TeamA	Tuesday	239	239	Spence	Thomas
TeamB	Monday	153	153	Nolan	Brenda

Join Condition
t.Manager = m.MemberID

Figure 6-7. *Joining Member and Team to get additional information about a team's manager*

Now we will look at how to retrieve information involving both relationship types.

Process Approach

The information from the join shown in Figure 6-6 is not particularly helpful. We have the managers' IDs, but it would be more useful to have their names as well. We need another join. First, we'll have a look at what Access will do by default if you add both the Member and Team tables onto the query design interface. This is shown in Figure 6-8.

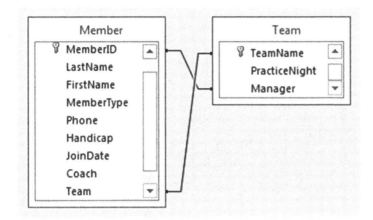

Figure 6-8. *Default joins in Access if Member and Team are added to diagrammatic query interface*

A look at the SQL for the query in Figure 6-8 reveals it is joining the tables like this:

```
SELECT *
FROM Member m   INNER JOIN Team t
ON t.TeamName = m.Team AND m.MemberID = t.Manager;
```

Can you figure out what question this query is answering? The output is shown in Figure 6-9.

One join with complex condition.
m.Team = t.TeamName AND t.Manager = m.MemberID

Figure 6-9. *Output for the default Access join in Figure 6-8*

To understand what is happening with the preceding join it is useful to consider the Cartesian product of Member and Team. The Cartesian product gives us every combination of rows from each table. The join condition says show only rows where the MemberID is the same as the Manager and where Team and TeamName are the same. In everyday language, this amounts to "Show me the members who manage the team they are in." For our data, that is just the single row for Brenda Nolan we see in Figure 6-9.

So, how do we construct a query that will show us member names, their teams, and the names of the teams' managers? The query that follows will provide the information about the members, their teams, and the managers' IDs (t.Manager); however, it does not provide the managers' names:

```
SELECT m.MemberID, m.LastName, m.FirstName, t.TeamName, t.Manager
FROM Member m INNER JOIN Team t ON m.Team = t.TeamName;
```

What we need to do is to take the result of the preceding join and join that to a *second* copy of the Member table (m2) to retrieve the names of the managers. We want the join condition to be that t.Manager = m2.MemberID so we get the names of the manager. Figure 6-10 shows a diagrammatic representation and the output of the two joins.

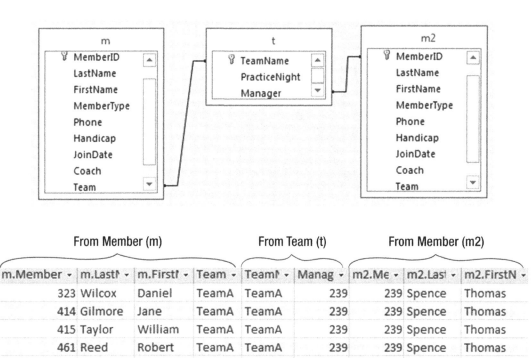

Figure 6-10. *Two joins and two copies of Member table to include names of team managers*

The first join gives us the member information from the first copy of the Member table and the information from the Team table for that member; the second join gives us the name of the team manager from the second copy of the Member table. The SQL for the two joins is:

```
SELECT *
FROM (Member m INNER JOIN Team t ON m.Team = t.TeamName)
    INNER JOIN Member m2 ON t.Manager = m2.MemberID;
```

You might find it instructive to compare this latest query and output with the query involving a single join between the Member and Team tables shown in Figures 6-8 and 6-9.

We are now in position to generate a variety of reports about teams and their members. Figure 6-11 shows a report based on the preceding query and its output, shown in Figure 6-10.

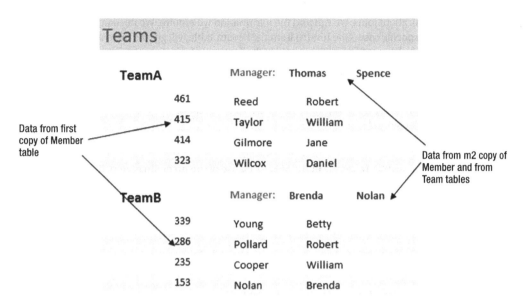

Figure 6-11. *A report based on the query shown in Figure 6-10*

The report has been grouped by team, with the team and manager information (from the Team table and m2 copy of the Member table) in a group header. The members of the team (from the first copy m of the Member table) are in the detail part of the report.

Outcome Approach

We will now look at an alternative way to construct a query to retrieve all the information about a team (members' names, team name, and manager's name) for a report like the one in Figure 6-11. I find the idea of two joins quite intuitive, but other people prefer a different approach.

I have reproduced the two tables in Figure 6-12. Now, without thinking about joins, let's see how we can pick a member and find out what team he or she is on and who the manager is for that team.

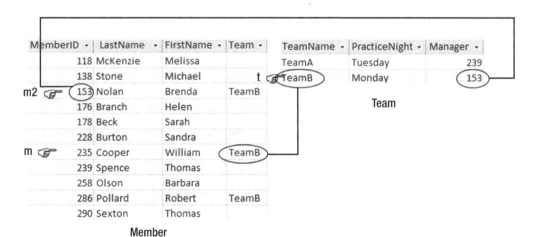

Figure 6-12. *Finding a team member (William Cooper), his team's name, and the name of the team's manager*

Without needing to think about joins, we can find the information we require. We need information from three rows. Let's look at one specific case. One row (m) from the Member table will give us the name of a member (William Cooper in Figure 6-12). We need to find the row (t) in the Team table for his team (m.Team = t.TeamName). Then we need another row in the Member table (m2) for the manager of the team (t.Manager = m2.MemberID).

With help from Figure 6-12 we can construct the following SQL:

```
SELECT m.LastName, m.FirstName, m.Team, m2.LastName, m2.FirstName
FROM Member m, Team t, Member m2
WHERE m.Team = t.TeamName AND t.Manager = m2.MemberID
```

We could replace m.Team with t.TeamName in the SELECT clause of the preceding query if we wish.

The preceding query is equivalent to the query with the two joins. The FROM clause is the Cartesian product of the three tables. The WHERE clause provides the join condition for the join between Member (m) and Team (t) on m.Team = t.TeamName and the join condition for the join between Team and another copy of Member (m2) on t.Manager = m2.MemberID.

Business Rules

The data model from Figure 6-4 is redisplayed below as Figure 6-13.

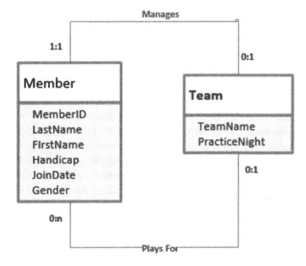

Figure 6-13. *Two relationships between the Member and Team classes*

Members can belong to teams, and members can manage teams. When we implement these relationships with foreign keys, the constraints that are placed on the data are quite simple. A member can only be on a team that exists in the Team table, and a team can be managed only by someone in the Member table.

Other constraints are likely to apply in various situations. For example, we might have the additional constraints that a team can have no more than four members or that the manager must be a member of the team (or not). These types of constraints are commonly referred to as *business rules*. The data model in Figure 6-13 might underpin a database for two different golf clubs. While the basic integrity rules will apply for both clubs (e.g., a member cannot be on a team that doesn't exist), each club might have different rules about sizes of teams and who can manage them. The foreign key constraints are not sufficient to enforce such business rules.

Relational database products will usually provide some way to enforce business rules. Large systems such as SQL Server and Oracle provide *triggers*. Triggers are actions that take place when a specified event occurs (for example, when inserting or updating a record). The trigger will reject any changes that do not obey the rules. In Access and other products, it is not possible to apply such constraints to the tables themselves. However, you can attach macros to input forms. These macros will check the data on the form before it is committed to the database. The issue with this approach is that there is no such checking if a user bypasses the form and enters data directly into a table with (for example) an SQL update command.

We won't look in detail at how business rules are implemented in different products, but we will look at how queries can help find any instances where the constraints are not satisfied. Although this is finding the problem after it has occurred, variations of these queries would form a basis for any trigger or macro that you would need to write to enforce the constraints.

Let's look at finding teams whose managers are not members of the team. My mind often goes blank when faced with a query like this, and in that case, I always take an outcome approach. This means picturing the tables involved and imagining the type of instance I am seeking. Have a look at Figure 6-14.

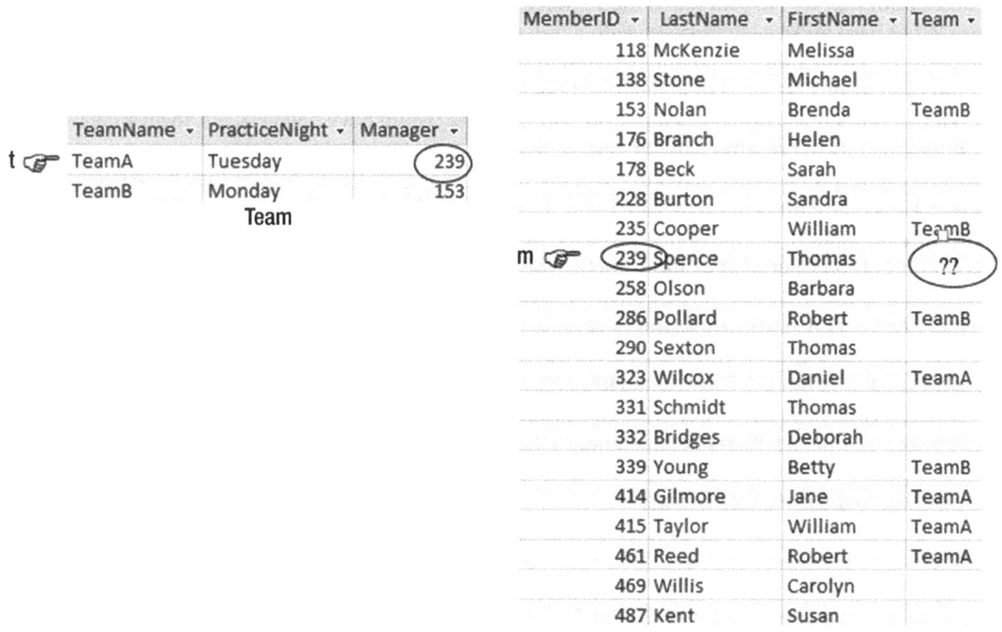

TeamName	PracticeNight	Manager
TeamA	Tuesday	239
TeamB	Monday	153

Team

MemberID	LastName	FirstName	Team
118	McKenzie	Melissa	
138	Stone	Michael	
153	Nolan	Brenda	TeamB
176	Branch	Helen	
178	Beck	Sarah	
228	Burton	Sandra	
235	Cooper	William	TeamB
239	Spence	Thomas	
258	Olson	Barbara	
286	Pollard	Robert	TeamB
290	Sexton	Thomas	
323	Wilcox	Daniel	TeamA
331	Schmidt	Thomas	
332	Bridges	Deborah	
339	Young	Betty	TeamB
414	Gilmore	Jane	TeamA
415	Taylor	William	TeamA
461	Reed	Robert	TeamA
469	Willis	Carolyn	
487	Kent	Susan	

Figure 6-14. *Finding teams whose managers are not members of the team*

In Figure 6-14, we see in the `Team` table that TeamA's manager is 239, and we can see in the `Member` table that member 239 is not a member of any team. If we had a constraint that managers must belong to the team, TeamA would not obey it.

To find all teams like this, we would say:

> *Find the team names from all the rows (`t`) in the `Team` table where the matching row (`m`) in the `Member` table for the team manager (i.e., `t.Manager = m.MemberID`) has a team (`m.team`) that is either empty or different from the team in the `Team` table (`m.Team <> t.TeamName`).*

The equivalent SQL is shown here:

```
SELECT t.teamname
FROM Member m, Team t
WHERE m.MemberID = t.Manager
AND (m.Team <> t.Teamname OR m.Team IS NULL)
```

The middle two lines are equivalent to a join between the two tables on m.MemberID = t.Manager, and the final line finds those managers who are on a different team or not on a team at all. The following query will produce an equivalent output but uses the inner join notation:

```
SELECT t.teamname
FROM Member m INNER JOIN Team t ON m.MemberID = t.Manager
WHERE m.Team <> t.Teamname OR m.Team IS NULL
```

Just a note about why we have included the IS NULL condition in the two queries: You might remember from Chapter 2 that if we make a comparison with a null value, the result is neither true nor false. If we want to find managers who aren't in a team, we need to specifically include that possibility in our query. Had the requirement been just that a manager must not belong to a different team, we could have left out the checking of null values, because a manager with no team would have been OK. As always, clearly understanding what you are actually trying to find is the most important part of specifying a query.

The two preceding queries will find teams with incorrect managers, but only after they have been added to the database. How do we prevent them from being added in the first place? The solution depends on the database implementation. Before changes to data are finally committed to a database, they are usually recorded in a buffer of some sort. For example, in SQL Server, the records being updated or added are kept in a temporary table called inserted. If we add or update some records to the Team table, a temporary table (inserted) that has the same structure as the Team table is created to hold the new or updated records temporarily. We want to perform a query to check if any new records about to be added to the Team table have managers that don't obey the constraint. However, instead of looking at the Team table, we want to look at the records in the temporary inserted table and count how many of those are invalid.

The following SQL query, which is very similar to the previous two queries, will count how many of the rows in the inserted buffer for the Team table have managers that do not obey the business rule about managers belonging to the team they manage:

```
SELECT COUNT(*)
FROM Member m INNER JOIN inserted i ON m.MemberID = i.Manager
WHERE m.Team <> i.Teamname OR m.Team IS NULL
```

If this count is not zero then there are rows that are about to be inserted that do not obey the rules. In that case we want to *rollback* the insertion so the rows do not get committed to the Team table. The following SQL statement would be included in a trigger in SQL. The trigger would need to be assigned to run on updating or inserting rows in the Team table.

```
IF
    (SELECT COUNT(*)
     FROM Member m INNER JOIN inserted i ON m.MemberID = i.Manager
     WHERE m.Team <> i.Teamname OR m.Team IS NULL)
    <> 0)
BEGIN
    Rollback Tran
END
```

This is a bit of a crude approach, because if any of the new records are incorrect, the whole lot gets rejected. You will need to consult the documentation for your database product to see how to develop triggers that work efficiently, but the idea of using a query to check the validity of new records is a common one.

In Access, the checking is done at the interface level, usually on a form. Instead of checking the `inserted` table as in the previous query, we would create a macro with a similar query to investigate the values of fields on the form before committing them to the database.

Summary

There can be more than one relationship between tables. For example, "a member may belong to a team" is one relationship. "A team has a club member who is the manager" is another relationship. Finding information about a member's team (including the manager's ID) requires a join between `Member` and `Team`. If we want to also find the name of the manager, we need to join that result to a second copy of the `Member` table, like this:

```
SELECT * FROM
(Member m INNER JOIN Team t ON m.Team = t.TeamName)
INNER JOIN Member m2 ON t.Manager = m2.MemberID
```

There can be quite complex business rules or constraints involving the relationships between tables. For example, we might require that the manager be a member of the team he or she manages, or that a manager should not be a member of any team, or that a team must have fewer than six members. These often require the use of triggers. The types of queries discussed in this chapter will be helpful in formulating the code required in triggers.

CHAPTER 7

■ ■ ■

Set Operations

One of the great strengths of relational database theory is that the tables (or, more formally, the relations) are made up of *distinct* rows and so can be considered a *set*. We can then use set operations to help with combining and extracting specific information. The types of questions that set operations help with are those such as "which people are in both these sets?" or "which people are in this set but not that one?"

In Appendix 2 you can find some formal notation that is helpful with managing set operations. In this chapter we will keep formalities to a minimum, but the symbols for the set operations are a useful shorthand. Table 7-1 shows the four set operations we will look at along with their common symbols and the associated SQL keyword (for those that have them).

Table 7-1. *Four Set Operations and Their Symbols*

Operation	Symbol	SQL Keyword
Union	∪	UNION
Intersection	∩	INTERSECT
Difference	−	EXCEPT
Division	÷	

Not all implementations of SQL support all the keywords in Table 7-1, so we will look at alternative ways to achieve the same result when the keywords are not available.

Overview of Basic Set Operations

We will look at each of the set operations in turn, but so that you know where we are heading, I'll just give a very quick overview of the three most common operations: union, intersection, and difference. Imagine we have membership tables from two golf clubs. We might want to do the following:

- Determine who is in both clubs.

- Form a large list that combines all the members.

- Find out who is in one club but not the other.

The basic set operations allow us to carry out all these tasks.

Let's assume that the each club keeps the names of its members in a table. The two tables have exactly the same columns (more about this in the next section) and are shown in Figure 7-1. (OK, they are very small clubs!)

© Clare Churcher 2016

C. Churcher, *Beginning SQL Queries*, DOI 10.1007/978-1-4842-1955-3_7

LastName ▾	FirstName ▾
Cooper	William
Gilmore	Jane
Kent	Susan
McKenzie	Melissa
Nolan	Brenda
Olson	Barbara
Pollard	Robert

ClubA

LastName ▾	FirstName ▾
Olson	Barbara
Pollard	Robert
Reed	Robert
Schmidt	Thomas
Sexton	Thomas

ClubB

Figure 7-1. *Two tables of member names*

The basic set operations on these two tables are summarized in Figure 7-2. The images of two club tables have been overlaid so that the members in common are superimposed. ClubA is the top table in each picture. For each section of Figure 7-2, the box shows the result of the set operation.

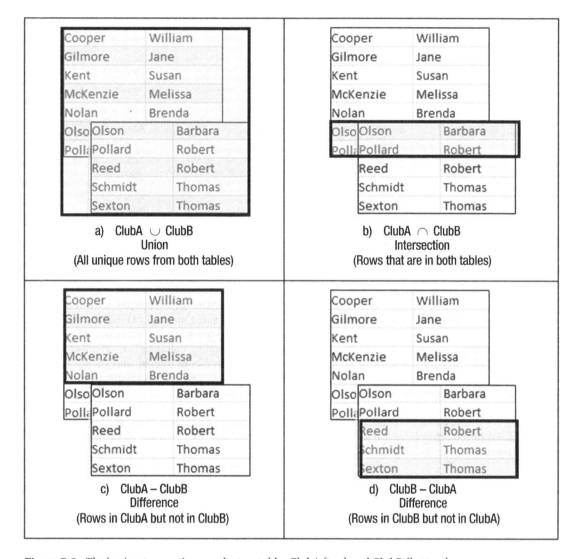

Figure 7-2. *The basic set operations on the two tables ClubA (top) and ClubB (bottom)*

The union operator (top left in Figure 7-2) shows all the names from each table (with duplicates removed). The intersect operator (top right) returns the two rows that appear in both tables. The difference operators (bottom) return those rows that are found in one club but not the other.

Union-Compatible Tables

The set operations union, intersection, and difference operate between two sets of rows. It does not make any sense to try to compare rows in tables that have very different structures, such as those in Figure 7-3.

MemberID	LastName	FirstName	Handicap
118	McKenzie	Melissa	30
138	Stone	Michael	30
153	Nolan	Brenda	11
176	Branch	Helen	
178	Beck	Sarah	
228	Burton	Sandra	26
235	Cooper	William	14
239	Spence	Thomas	10
258	Olson	Barbara	16

MemberID	TourID	Year
118	24	2014
228	24	2015
228	25	2015
228	36	2015
235	38	2013
235	38	2015
235	40	2014
235	40	2015
239	25	2015
239	40	2013
258	24	2014

Member Table

Entry Table

Figure 7-3. *It makes no sense to try to compare rows from tables with different structures*

So what determines whether two sets of rows can be compared using the set operations union, intersection, and difference? Formally, the two sets must have the same number of columns, and each column must have the same domain. Strictly speaking, a *domain* is a set of possible values. However, in practice, the requirement for set operations is that the corresponding columns (i.e., in order from left to right) in each set of rows have the same types – both character, both integer, and so on.[1] The names of the columns do not need to be the same. Tables that meet these requirements are referred to as being *union compatible*, although the requirement is necessary for the intersection and difference operations as well.

Figure 7-4 shows a pair of tables that are union compatible. Even though the names of the columns are different, they have the same number of columns, and the corresponding columns have the same types.

---

[1]In formal relational theory, the attributes of a relation have no order but rather are referenced by their names. The ordering of columns in tables is how implementations of SQL determine union compatibility.

MemberID ▾	LastName ▾	FirstName ▾	Handicap ▾	MemberType ▾
176	Branch	Helen		Social
178	Beck	Sarah		Social
228	Burton	Sandra	26	Junior
235	Cooper	William	14	Senior
239	Spence	Thomas	10	Senior
258	Olson	Barbara	16	Senior
286	Pollard	Robert	19	Junior
290	Sexton	Thomas	26	Senior

ClubA

RegNum ▾	FamilyName ▾	Name ▾	Handicap ▾	Grade ▾
239	Spence	Thomas	10	Senior
258	Olson	Barbara	16	Senior
286	Pollard	Robert	19	Junior
290	Sexton	Thomas	26	Senior
323	Wilcox	Daniel	3	Senior
331	Schmidt	Thomas	25	Senior
332	Bridges	Deborah	12	Senior
339	Young	Betty	21	Senior

ClubB

Figure 7-4. *Union-compatible tables, even though column names are different*

Figure 7-5 has two tables with the same column names, but they are not union compatible because the order of the columns is such that the fourth column has a number type in the top table and a character type in the bottom, and vice versa for the last column.

MemberID ▾	LastName ▾	FirstName ▾	Handicap ▾	MemberType ▾
176	Branch	Helen		Social
178	Beck	Sarah		Social
228	Burton	Sandra	26	Junior
235	Cooper	William	14	Senior
239	Spence	Thomas	10	Senior
258	Olson	Barbara	16	Senior
286	Pollard	Robert	19	Junior
290	Sexton	Thomas	26	Senior

ClubC

MemberID ▾	LastName ▾	FirstName ▾	MemberType ▾	Handicap ▾
239	Spence	Thomas	Senior	10
258	Olson	Barbara	Senior	16
286	Pollard	Robert	Junior	19
290	Sexton	Thomas	Senior	26
323	Wilcox	Daniel	Senior	3
331	Schmidt	Thomas	Senior	25
332	Bridges	Deborah	Senior	12
339	Young	Betty	Senior	21

ClubD

Figure 7-5. *Tables that are not union compatible*

Different implementations of SQL may interpret the strictness of this requirement for the "sameness" of domains or types differently. Strictly speaking, two fields defined as CHAR(10) and CHAR(12) have different domains, but many implementations of SQL will allow these to be regarded as the same for the purposes of set operations. Some implementations will also convert numbers into characters to enable set operations to be carried out. I find this particularly scary and don't recommend you let your application make these sorts of decisions for you. The following sections demonstrate how you can use SQL to make your tables union compatible.

Ensuring Union Compatibility

When tables are not union compatible, you can often remedy the incompatibility in the SELECT clauses.

For the pair of tables in Figure 7-5, if we just select the columns as follows the order of the columns will prevent the returned rows from being union compatible:

```
SELECT * FROM ClubC;
SELECT * FROM ClubD;
```

However, we can specify the order of the columns in the SELECT clause:

```
SELECT MemberID, LastName, FirstName, Handicap, MemberType FROM ClubC;
SELECT MemberID, LastName, FirstName, Handicap, MemberType FROM ClubD;
```

The two sets of rows returned from these queries are now union compatible.

Another incompatibility problem occurs when the types of the columns have been declared as different types in the original design of the tables. For example, the ClubC table may have the Handicap field declared as an INT, whereas the ClubD table may have (unwisely) stored the Handicap values in a CHAR field. (Recall from Chapter 2 that if we store values in a character or text field then they will order alphabetically, and we will not be able to perform functions such as average on them.) As mentioned earlier, different implementations of SQL will treat these different types in a variety of ways. Many will try to convert the numbers to text or vice versa. You can take control of these conversions yourself (which is probably a good idea) by using type-conversion functions.

For example, in SQL Server, the expression Convert(INT, Handicap) would take a text value in the Handicap field ("14") and convert it to an integer value (14). (If the value in the Handicap field wasn't able to be converted to an integer then an error would occur.) If the Handicap field in the ClubD table were a CHAR type then we could use the conversion function in the SELECT clause. The two sets of rows returned by the following queries will now be union compatible:

```
SELECT MemberID, LastName, FirstName, Handicap FROM ClubC;
SELECT MemberID, LastName, FirstName, Convert(INT, Handicap) FROM ClubD;
```

Union

Union allows us to produce output consisting of all the unique rows from two union-compatible sets of rows. To carry out a union in SQL, we need to first retrieve two sets of rows using two SELECT clauses and then combine the two sets with the UNION keyword. The following SQL shows the union of all the rows from the two union-compatible tables (ClubA and ClubB) shown in Figure 7-4.

```
SELECT * FROM ClubA
UNION
SELECT * FROM ClubB;
```

The resulting table will include all the rows from both tables with no duplicates, so you will see only one row each for Barbara Olson, Robert Pollard, and Thomas Sexton, as shown in Figure 7-6. If you wish to retain the duplicates for some reason, you can use the key phrase UNION ALL.

MemberID ▾	LastName ▾	FirstName ▾	Handicap ▾	MemberType ▾
176	Branch	Helen		Social
178	Beck	Sarah		Social
228	Burton	Sandra	26	Junior
235	Cooper	William	14	Senior
239	Spence	Thomas	10	Senior
258	Olson	Barbara	16	Senior
286	Pollard	Robert	19	Junior
290	Sexton	Thomas	26	Senior
323	Wilcox	Daniel	3	Senior
331	Schmidt	Thomas	25	Senior
332	Bridges	Deborah	12	Senior
339	Young	Betty	21	Senior

Figure 7-6. *Union of ClubA and ClubB with no duplicate rows*

As union-compatible tables do not need to have the same column names, the names of the columns in the resulting virtual table will usually be from one of the tables. In the example in Figure 7-6, the column names are the same as the first table mentioned in the union query.

It does not matter for the union operator in which order the two tables are specified. The query that follows will return the same rows as the previous query did. The rows may appear in a different order, and the displayed names of the columns may change, but the data will be the same.

```
SELECT * FROM ClubB
UNION
SELECT * FROM ClubA;
```

Selecting the Appropriate Columns

When using the union operator you need to think carefully about what it is you actually want. The examples with the clubs are rather contrived (as you have no doubt noticed). It is very unlikely that two clubs would have members with the same ID numbers and identical membership types. A more likely scenario is that if Barbara Olson did belong to two clubs, she would have different data in each club table. In the ClubA table, she might be a Senior with a value of 258 for MemberID. In the ClubB table, she might be an Associate with a value of 4573 for RegNum. If we do the union shown in Figure 7-6, where we select all the columns from each table, the two rows for Barbara will be different, and so both will appear in the result, as in Figure 7-7.

MemberID ▾	LastName ▾	FirstName ▾	MemberType ▾	Handicap ▾
176	Branch	Helen	Social	
178	Beck	Sarah		
228	Burton	Sandra	Junior	26
235	Cooper	William	Senior	14
239	Spence	Thomas	Senior	10
258	Olson	Barbara	Senior	16
4573	Olson	Barbara	Associate	16
286	Pollard	Robert	Junior	19

Figure 7-7. *Two records appear for Barbara Olson in the union because the rows are different*

We need to consider what we really want from such a union. If we need a list of names for a joint Christmas party for the two clubs, then we don't want everyone listed twice. The way to avoid duplicates is to project just the names from each table before carrying out the union:

```
SELECT FamilyName, Name FROM ClubA
UNION
SELECT LastName, FirstName FROM ClubB;
```

With this query the two rows for Barbara will be the same and will only appear in the union once, as in Figure 7-8.

LastName ▾	FirstName ▾
Branch	Helen
Beck	Sarah
Burton	Sandra
Cooper	William
Spence	Thomas
Olson	Barbara
Pollard	Robert

Figure 7-8. *Only one row appears for Barbara Olson if only the name columns are in the union*

There is, of course, a serious issue with this last query. There may be two Barbara Olsons, one in each club, and now only one nametag will be printed for the pair of them. Sadly, real data is fraught with these sorts of problems. With any luck there will be some universal national golf association number that might sort this out, but if not you just need to be alert. The intersection operation, discussed in the next section, would produce the names that appear in both club lists, and a manual sanity check could be carried out.

Uses of Union

The main use for union is combining data from two or more tables, as we have been doing in the previous sections. For example, if the tournament entry data for different months had been stored in separate tables (not a great design decision!), we could use several union operations to combine the data for the whole year.

It is also possible to combine two sets of rows from the one table. Say we wanted to know how many people have entered either tournament 24 or tournament 36 from the Entry table in Figure 7-9.

MemberID ▾	TourID ▾	Year ▾
118	24	2014
228	24	2015
258	24	2014
286	24	2013
286	24	2014
286	24	2015
415	24	2015
228	25	2015
239	25	2015
415	25	2013
228	36	2015
415	36	2014
415	36	2015
235	38	2013
235	38	2015

Figure 7-9. *Entry table*

We could try selecting the rows for members entering tournament 24 and the rows for members entering tournament 36, and take the union. How many rows will we get if we perform the following query?

```
SELECT * FROM Entry WHERE TourID = 24
UNION
SELECT * FROM Entry WHERE TourID = 36;
```

We will get ten rows from this query, one for every row with a 24 or a 36. Because we have retained the TourID and Year columns, the rows we have selected are all different and so will all appear in the result of the union. The query is actually returning all the distinct *entries* into tournaments 24 and 36 rather than all the distinct *members* who have entered the two tournaments. The flowing query takes the union of just the IDs for the two tournaments:

```
SELECT MemberID FROM Entry WHERE TourID = 24
UNION
SELECT MemberID FROM Entry WHERE TourID = 36;
```

Now we will get the five IDs (118, 228, 258, 286, 415) that are the unique IDs for those entering one or the other of the tournaments.

There is a much simpler way of retrieving those who have entered either tournament 24 or 36. We can simply include an OR in the WHERE clause:

```
SELECT MemberID FROM Tournament
WHERE TourID = 24 OR TourID = 36;
```

How many rows will the preceding query return? Again, it will return ten rows–each of the rows with a 24 or a 36 in the TourID column. To get the five unique IDs, we need to add the DISTINCT keyword in the SELECT clause.

Union and Full Outer Joins

In Chapter 3 we looked at different join operations: inner joins, left and right outer joins, and full outer joins. Some products (e.g., Microsoft Access 2013) do not support the FULL OUTER JOIN keyword; however, we can perform an equivalent query using the UNION keyword.

To recap, let's review the different types of join we can carry out between the Member table (just a very little one!) and the Type table shown in Figure 7-10.

MemberID ▾	LastName ▾	FirstName ▾	MemberType ▾
118	McKenzie	Melissa	Junior
138	Stone	Michael	Senior
153	Nolan	Brenda	Senior
176	Branch	Helen	
178	Beck	Sarah	Social

Type ▾	Fee ▾
Associate	60
Junior	150
Senior	300
Social	50

Member   Type

Figure 7-10. *The (small) Member and Type tables*

Figure 7-11 shows the inner join between the two tables, with join condition MemberType = Type. We do not get a row for Helen Branch because she has no value in MemberType and so the join condition will never be true for her. This may be a problem if someone looking at the table in Figure 7-11 assumes it is showing all members.

MemberID ▾	LastName ▾	FirstName ▾	MemberType ▾	Type ▾	Fee ▾
118	McKenzie	Melissa	Junior	Junior	150
138	Stone	Michael	Senior	Senior	300
153	Nolan	Brenda	Senior	Senior	300
178	Beck	Sarah	Social	Social	50

Figure 7-11. *The inner join between Member and Type on MemberType = Type*

Now we will look at the outer joins. The left outer join ensures that we see all the rows from the left-hand table (Member); the right outer join gives us all rows from the right-hand table (Type); and the full outer join gives us all rows from both tables. These outer joins, all with join condition MemberType = Type, are shown in Figure 7-12.

MemberID ▾	LastName ▾	FirstName ▾	MemberType ▾	Type ▾	Fee ▾
118	McKenzie	Melissa	Junior	Junior	150
138	Stone	Michael	Senior	Senior	300
153	Nolan	Brenda	Senior	Senior	300
176	Branch	Helen			
178	Beck	Sarah	Social	Social	50

Member Left Join Type

MemberID ▾↑	LastName ▾	FirstName ▾	MemberType ▾	Type ▾	Fee ▾
				Associate	60
118	McKenzie	Melissa	Junior	Junior	150
138	Stone	Michael	Senior	Senior	300
153	Nolan	Brenda	Senior	Senior	300
178	Beck	Sarah	Social	Social	50

Member Right Join Type

MemberID ▾↑	LastName ▾	FirstName ▾	MemberType ▾	Type ▾	Fee ▾
				Associate	60
118	McKenzie	Melissa	Junior	Junior	150
138	Stone	Michael	Senior	Senior	300
153	Nolan	Brenda	Senior	Senior	300
176	Branch	Helen			
178	Beck	Sarah	Social	Social	50

Member Full Join Type

Figure 7-12. *Three outer joins between Member and Type on MemberType = Type*

Figure 7-12 shows that, in this case, the full outer join consists of the unique rows from each of the other two outer joins; that is, a union. If your SQL implementation does not explicitly support a full outer join, you can always achieve the same result with the following query:

```
SELECT * FROM Member LEFT JOIN Type ON MemberType = Type
UNION
SELECT * FROM Member RIGHT JOIN Type ON MemberType = Type;
```

Intersection

If you take the intersection of two union-compatible tables, you will retrieve those rows that are found in both tables. Figure 7-13 reproduces the two tables, ClubA and ClubB, from Figure 7-4. We can see that there are four rows that are identical in both tables.

MemberID ▾	LastName ▾	FirstName ▾	Handicap ▾	MemberType ▾
176	Branch	Helen		Social
178	Beck	Sarah		Social
228	Burton	Sandra	26	Junior
235	Cooper	William	14	Senior
239	Spence	Thomas	10	Senior
258	Olson	Barbara	16	Senior
286	Pollard	Robert	19	Junior
290	Sexton	Thomas	26	Senior

ClubA

RegNum ▾	FamilyName ▾	Name ▾	Handicap ▾	Grade ▾
239	Spence	Thomas	10	Senior
258	Olson	Barbara	16	Senior
286	Pollard	Robert	19	Junior
290	Sexton	Thomas	26	Senior
323	Wilcox	Daniel	3	Senior
331	Schmidt	Thomas	25	Senior
332	Bridges	Deborah	12	Senior
339	Young	Betty	21	Senior

ClubB

Figure 7-13. *The rows in the intersection between the ClubA and ClubB tables*

The keyword for the intersection operator in SQL is INTERSECT. The expression to retrieve the four rows common to both tables (i.e., for members Spence, Olson, Pollard, and Sexton) is as follows:

```
SELECT * FROM ClubA
INTERSECT
SELECT * FROM ClubB;
```

As with the union operator, the two sets of rows must be union compatible; that is, they must have the same number of columns, and the corresponding columns must have the same domains. This may mean projecting the appropriate columns from the base tables in the same way as described in the "Selecting the Appropriate Columns" section earlier in this chapter. It makes no difference which of the tables we mention first in the query, as the rows returned by the intersection will be the same regardless of the order of the tables.

Uses of Intersection

A common use of the intersection operation is the one shown in Figure 7-13: finding common rows in two tables with similar information. Another very common use of intersection is answering questions that include the word *both*. A typical example is "Which members have entered *both* tournaments 36 and 38?" The Entry table is reproduced in Figure 7-14.

MemberID ▾	TourID ▾	Year ▾
118	24	2014
228	24	2015
258	24	2014
286	24	2013
286	24	2014
286	24	2015
415	24	2015
228	25	2015
239	25	2015
415	25	2013
228	36	2015
415	36	2014
415	36	2015
235	38	2013
235	38	2015
258	38	2014
415	38	2013
415	38	2015
235	40	2014
235	40	2015
239	40	2013
415	40	2013
415	40	2014
415	40	2015

Figure 7-14. *The Entry table*

What will be returned if we retrieve the rows for each tournament and take the intersection as in the following query?

```
SELECT * FROM Entry WHERE TourID = 36
INTERSECT
SELECT * FROM Entry WHERE TourID = 38;
```

There will be no rows returned. Figure 7-15 will help you understand why.

MemberID ▾	TourID ◂	Year ▾
228	36	2015
415	36	2014
415	36	2015

MemberID ▾	TourID ◂	Year ▾
235	38	2013
235	38	2015
258	38	2014
415	38	2013
415	38	2015

SELECT * FROM Entry
WHERE TourID = 36

SELECT * FROM Entry
WHERE TourID = 38

Figure 7-15. *Two queries have no rows in common, so no rows result from intersection*

The two queries will never have any rows in common because one will always have 36 in the TourID column while the other will always have 38. Essentially, the query we were trying to carry out was to find all the entries for tournament 36 that are also entries for tournament 38. The result, given the way we are managing entries, is none.

To retrieve the members who are in common in the two sets of rows in Figure 7-15, we must retrieve just the MemberID column before carrying out the intersection, as in the query here:

```
SELECT MemberID FROM Entry WHERE Tourid = 36
INTERSECT
SELECT MemberID FROM Entry WHERE Tourid = 38;
```

This query is illustrated in Figure 7-16. As with a union, the result of the intersection operation returns unique rows.

MemberID ▾
228
415
415

∩

MemberID ▾
235
235
258
415
415

=

MemberID ▾
415

SELECT MemberID
FROM Entry
WHERE TourID=36;

SELECT MemberID
FROM Entry
WHERE TourID=38

The intersection

Figure 7-16. *Using intersection to find members entered in both tournaments 36 and 38*

Suppose we now want to find the names of the members. From a process point of view, we can take the result of the intersection and join it with the Member table to get the names, as shown in Figure 7-17.

MemberID ▾	LastName ▾	FirstName ▾	MemberID ▾	MemberID ▾	LastName ▾	FirstName ▾
118	McKenzie	Melissa	415	415 Taylor		William
138	Stone	Michael				
153	Nolan	Brenda				
176	Branch	Helen				
178	Beck	Sarah				
228	Burton	Sandra				
235	Cooper	William				
239	Spence	Thomas				
258	Olson	Barbara				
286	Pollard	Robert				
290	Sexton	Thomas				
323	Wilcox	Daniel				
331	Schmidt	Thomas				
332	Bridges	Deborah				
339	Young	Betty				
414	Gilmore	Jane				
415	Taylor	William				
461	Reed	Robert				
469	Willis	Carolyn				
487	Kent	Susan				

Member Table   Intersection   Member Inner Join Intersection

Figure 7-17. *Joining the intersection with the Member table to find the names*

So what does the SQL look like to first do the intersection and then join with the Member table? The following is a good first attempt, but unfortunately will not work:

```
--Will not work
SELECT LastName, FirstName
FROM Member m INNER JOIN
    (SELECT e1.MemberID FROM Entry e1 WHERE e1.TourID = 36
     INTERSECT
     SELECT e2.MemberID FROM Entry e2 WHERE e2.TourID = 38)
ON m.MemberID = e1.MemberID;
```

The tables that only appear inside the inner query (the part in parentheses) are not able to be referenced by the outer query (the join). This is easily resolved by giving the nested part of the query an alias. In the same way we have given the Member table an alias by putting an m after Member in the FROM clause, we can give the whole inner query an alias of NewTable (as an example) by putting NewTable after the final parenthesis of the inner query. We can now refer to that alias in the join condition as shown in the query here:

```
SELECT LastName, FirstName
FROM Member m INNER JOIN
    (SELECT e1.MemberID FROM Entry e1 WHERE e1.TourID = 36
```

114

```
    INTERSECT
    SELECT e2.MemberID FROM Entry e2 WHERE e2.TourID = 36) NewTable
ON m.MemberID = NewTable.MemberID;
```

Another way to retrieve the names is to use a nested query. Here, the inner query retrieves the IDs that are in the intersection, and the outer query finds the corresponding names from the Member table.

```
SELECT LastName, FirstName
FROM Member
WHERE MemberID IN
    (SELECT MemberID FROM Entry WHERE TourID = 36
     INTERSECT
     SELECT MemberID FROM Entry WHERE TourID = 38);
```

The Importance of Projecting Appropriate Columns

It is important to think very carefully about which columns are included in the tables involved in an intersection operation. We saw in the previous section how the following query will return no rows:

```
SELECT * FROM Entry WHERE TourID = 36
INTERSECT
SELECT * FROM Entry WHERE TourID = 38;
```

The rows from the first query will always have 36 as the value of TourID and the rows from the second query will have 38. There will never be any rows in common. Retrieving just the MemberID in each of the queries solves this problem.

More interesting is that correctly projecting different columns can provide answers to quite different questions. How would you describe the rows returned by the following query?

```
SELECT MemberID, Year FROM Entry WHERE TourID = 25
INTERSECT
SELECT MemberID, Year FROM Entry WHERE TourID = 36;
```

The query is illustrated in Figure 7-18.

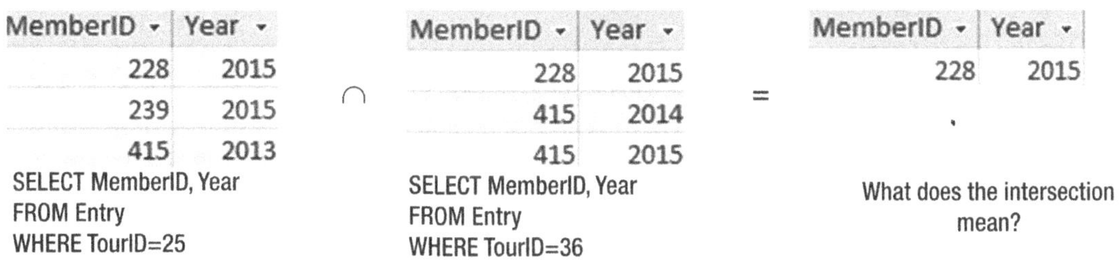

Figure 7-18. *What does the intersection mean?*

In Figure 7-18, we are finding all the members who entered tournaments 25 and 36 in the *same year*. This is why there is no entry for member 415 in the intersection: he entered tournament 25 in 2013 and tournament 36 in 2014 and 2015. Although his member ID appears in the two contributing tables, the corresponding rows are for different years. There is no row for member 415 that is the same in both tables.

As you can see, the choice of columns that are projected for the contributing tables is fundamental to what will appear in the intersection. It means there are many different questions that can be answered very elegantly, but it also means that you can easily get incorrect answers if you don't think the query through carefully.

Managing Without the INTERSECT Keyword

Not all implementations of SQL support intersection explicitly. However, we have other ways to perform the queries involving "both." Intersection is a process approach – we are saying what operations we need to carry out on the tables involved in the query. If we don't succeed with this approach then we can try the outcome approach. This involves figuring out some possible answers by inspecting the tables and not worrying about operations such as intersections and joins. In Figure 7-19 we imagine two fingers traversing the rows of the Entry table. We need to find two rows in the Entry table with the same MemberID: one with TourID = 36 and one with TourID = 38.

MemberID ▾	Year ▾	TourID ▾
415	2013	25
228	2015	36
e1 ☞ (415)	2014	[36]
415	2015	36
235	2013	38
235	2015	38
258	2014	38
e2 ☞ (415)	2013	[38]
415	2015	38
235	2014	40

Figure 7-19. *Finding members who have entered both tournaments 36 and 38*

The situation that Figure 7-19 is depicting can be described as:

Return me the MemberID from a row e1 in the Entry table where TourID=36 if there is another row e2 in the Entry table that has the same MemberID and TourID=38 .

The SQL expression equivalent to this description and Figure 7-19 is:

```
SELECT DISTINCT e1.MemberID
FROM Entry e1, Entry e2
WHERE e1.MemberID = e2.MemberID
AND e1.TourID = 36 AND e2.TourID = 38;
```

What about the query to find the rows that appear in both the ClubA and ClubB tables? The club tables are redisplayed in Figure 7-20. To find the rows that are the same in both tables we need to check each of the values in the corresponding columns to ensure they are the same.

RegNum	FamilyName	Name
176	Branch	Helen
178	Beck	Sarah
228	Burton	Sandra
235	Cooper	William
239	Spence	Thomas
258	Olson	Barbara
286	Pollard	Robert

a ☞

ClubA

MemberID	LastName	FirstName
258	Olson	Barbara
286	Pollard	Robert
290	Sexton	Thomas
323	Wilcox	Daniel
331	Schmidt	Thomas
332	Bridges	Deborah
339	Young	Betty

b ☞

ClubB

Figure 7-20. *Finding the intersection between ClubA and ClubB*

The situation depicted in Figure 7-20 can be described as:

I will return row a from table ClubA if there is a row b in ClubB that has identical values in all the fields (i.e., a.RegNum = b.MemberID, a.FamilyName = b.LastName, and a.Name = b.FirstName).

The SQL for the intersection shown in Figure 7-20 is:

```
SELECT a.RegNum, a.FamilyName, a.Name
FROM ClubA a, ClubB b
WHERE a.RegNum = b.MemberID
AND a.FamilyName = b.LastName
AND a.Name = b.FirstName;
```

117

Difference

Taking the difference between two tables finds those rows that are in the first table but not the second and vice versa. For our two tiny clubs, I have reproduced the results of the difference operator in Figure 7-21.

Cooper	William			
Gilmore	Jane			
Kent	Susan			
McKenzie	Melissa			
Nolan	Brenda			
Olso	Olson	Barbara		
Poll	Pollard	Robert		
	Reed	Robert		
	Schmidt	Thomas		
	Sexton	Thomas		

ClubA – ClubB   ClubB – ClubA

Figure 7-21. *The difference operator finds rows in one table that do not appear in the other.*

The keyword in standard SQL for the difference operator is EXCEPT. Oracle differs from the ISO SQL standard, and from most other database systems, in its use of the keyword MINUS rather than EXCEPT.

As with the union and intersection operators, the tables involved in a difference operation must be union compatible. Unlike with the union and intersection operators, the order of the tables is important for the difference operator; the results for ClubA - ClubB are different from those for ClubB - ClubA (as shown in Figure 7-21).

The SQL for finding the names of people in the ClubA table that do not appear in the ClubB table is:

```
SELECT LastName, FirstName FROM ClubA
EXCEPT
SELECT LastName, FirstName FROM ClubB;
```

Uses of Difference

Whenever you have a query that has the word "not," you should consider the possibility that the difference operator will be useful. For example, how do we find members who have not entered tournament 25? Recall from Chapter 5 why the following query does not return those members who have not entered tournament 25:

```
SELECT MemberID FROM Entry
WHERE TourID <> 25;
```

The query above selects all the rows in the Entry table that are not for tournament 25. Essentially it finds a member who has entered any tournament other than 25 (although they could have entered 25 as well). Looking at Figure 7-15, we see that the query would return the row marked e1 for member 415 entering tournament 36 (TourID <> 25). However, two rows above, we see that member 415 has also entered tournament 25. It is difficult to think of a reason that you might ever want to use this query.

A process approach to this type of query is to use difference. We need to retrieve a set of the IDs of all members and another set of IDs for all the members who have entered tournament 25. We then want the difference; i.e., those IDs that are in the former set but not the latter.

Finding the set of all members who have entered tournament 25 is simple:

```
SELECT MemberID FROM Entry WHERE TourID = 25;
```

We might think a similar query will find us all the member IDs as well:

```
SELECT MemberID FROM Entry;
```

However, the preceding query only finds us a set of the members who have entered tournaments. To get a set of all member IDs, we need to query the Member table.

Figure 7-22 is an illustration of how the difference operator can be used to find the member IDs we require.

MemberID ▾	MemberID ▾	MemberID ▾
118		118
138		138
153		153
176		176
178		178
228		235
235		258
239	228	286
258	239	290
286	415	323
290		331
323		332
331		339
332		414
339		461
414		469
415		487
461		
469		
487		

A	B	A-B
IDs of all members	IDs of members entering 25	IDs of members who have not entered 25

Figure 7-22. *Members who have not entered tournament 25*

The SQL expression to retrieve the IDs of members who have not entered tournament 25 is as follows:

```
SELECT MemberID FROM Member
EXCEPT
SELECT MemberID FROM Entry WHERE TourID = 25;
```

As with intersection and union operations, it is important that we project the appropriate columns before we use the difference operator. In Figure 7-22, we have retrieved the IDs from the Member and Entry tables. If we want to include the names of the members, we can use one of the methods explained in the "Uses of Intersection" section earlier in this chapter.

However, in this difference example, we already had the names of the members in the Member table before we removed them to get the set of rows on the left side of Figure 7-22. It seems a bit perverse to remove the names and then put them back later. What is important is that the two sets of rows involved in the difference are union compatible; that is, the corresponding columns must have the same domains. Either both sets have just IDs or both sets have IDs and names. In the operation on the left side of Figure 7-22, we took the first option and removed the names from Member. We could have left the names in the Member table and added the names to the rows in the middle of Figure 7-22 by joining the Entry and Member tables, as shown in Figure 7-23. We could then take the difference between these two sets of rows.

MemberID ▾	LastName ▾	FirstName ▾
118	McKenzie	Melissa
138	Stone	Michael
153	Nolan	Brenda
176	Branch	Helen
178	Beck	Sarah
228	Burton	Sandra
235	Cooper	William
239	Spence	Thomas
258	Olson	Barbara
286	Pollard	Robert
290	Sexton	Thomas
323	Wilcox	Daniel
331	Schmidt	Thomas
332	Bridges	Deborah
339	Young	Betty
414	Gilmore	Jane
415	Taylor	William
461	Reed	Robert
469	Willis	Carolyn
487	Kent	Susan

m.Member ▾	LastName ▾	FirstName ▾	e.MemberI ▾	TourID ▾	Year ▾
239	Spence	Thomas	239	25	2015
228	Burton	Sandra	228	25	2015
415	Taylor	William	415	25	2013

A
IDs and Names from Member

B
Entry table joined with Member table
Then rows for tournament 25 selected
Then IDs and Names projected

Figure 7-23. *Including names of members in both sets of rows before taking the difference*

The SQL equivalent of the operations shown in Figure 7-23 is as follows:

```
SELECT MemberID, LastName, FirstName FROM Member
EXCEPT
SELECT m.MemberID, m.LastName, m.FirstName
FROM Entry e inner join Member m on e.MemberID = m.MemberID
WHERE TourID = 25;
```

Managing Without the EXCEPT Keyword

Not all versions of SQL support the EXCEPT (or MINUS) keyword. As always, there is usually another way to formulate a query. In Chapter 4, we looked at an outcome approach to answering questions involving the word *not*. Figure 7-24 reviews the thought processes used to find the names of members who have not entered tournament 25.

MemberID ▾	LastName ▾	FirstName ▾
118	McKenzie	Melissa
138	Stone	Michael
153	Nolan	Brenda
176	Branch	Helen
178	Beck	Sarah
228	Burton	Sandra
235	Cooper	William
239	Spence	Thomas
m ☞ 258	Olson	Barbara
286	Pollard	Robert
290	Sexton	Thomas
323	Wilcox	Daniel
331	Schmidt	Thomas
332	Bridges	Deborah
339	Young	Betty
414	Gilmore	Jane
415	Taylor	William
461	Reed	Robert
469	Willis	Carolyn
487	Kent	Susan

MemberID ▾	Year ▾	TourID ◂
118	2014	24
228	2015	24
e ☞ 258	2014	24 ?
286	2013	24
286	2014	24
286	2015	24
415	2015	24
228	2015	25
239	2015	25
415	2013	25
228	2015	36
415	2014	36
415	2015	36
235	2013	38
235	2015	38
258	2014	38 ?
415	2013	38
415	2015	38
235	2014	40
235	2015	40
239	2013	40
415	2013	40
415	2014	40
415	2015	40

Member table   **Entry table**

Figure 7-24. Deciding that member 258 has not entered tournament 25

The thought process behind Figure 7-24 is:

Write out the names from row m of the Member *table if there does not exist a row* e *in the* Entry *table for that member (i.e.,* m.MemberID = e.MemberID*) where* TourID=25*.*

The SQL reflecting Figure 7-24 is:

```
SELECT m.LastName, m.FirstName
FROM Member m
WHERE NOT EXISTS
    (SELECT * FROM Entry e
     WHERE e.MemberID = m.MemberID
     AND e.TourID = 25);
```

Which type of query should you use for questions involving the word *not*? The one using the process approach and the keyword EXCEPT or the one using the outcome approach with the keywords NOT EXISTS or NOT IN? Usually, I'd say it doesn't really matter, as your database engine will probably be smart enough to recognize them as being the same. However, the version of SQL Server I am using at the moment (2013) performs the query using NOT EXISTS more efficiently than the corresponding query using EXCEPT. You have to ask yourself whether you care! Queries on small databases are usually so quick that it really doesn't matter if they run a bit more slowly. However, if you have a lot of data, then everything changes. The efficiency of queries can become extremely important, and in that case, you will need to also consider other aspects of your database design, such as indexes. I'll talk a little more about this in Chapter 9.

Division

The last set operator we will look at in this chapter is division. Division is useful for queries that involve the word *all* or *every*. An example is "Which members have entered *every* tournament?" Standard SQL doesn't have a keyword for the divide operation, and it can be a little awkward to figure out the SQL for queries involving division.

In Appendix 2 you will find the formal algebraic notation for carrying out division and how to represent it using other operators if you need to. In the section "Universal Quantifier and SQL" in Appendix 2, you will also find an alternative way to carry out division-type queries using calculus (or outcome) expressions. Both these methods help you to construct SQL statements that are analogous to the division operator. In Chapter 8, we'll look at aggregates and see what I think is the simplest way of writing an SQL equivalent of the division operator.

For now we will look at what the division operator does and how to use it to answer different types of questions involving *every* and *all*.

The easiest way to understand the division operation is with an example. If we want to know which members have entered every tournament, we need two bits of information. First, we need information about the members and the tournaments they have entered, which we can get from the Entry table. We also need a list of all the tournaments, which needs to come from the Tournament table, as not all tournaments may be represented in the Entry table.

Figure 7-25 illustrates how division works. I've projected just the MemberID and TourID columns from the Entry table, and the TourID column from the Tournament table. It is important which columns you project, and I'll come back to that in a moment.

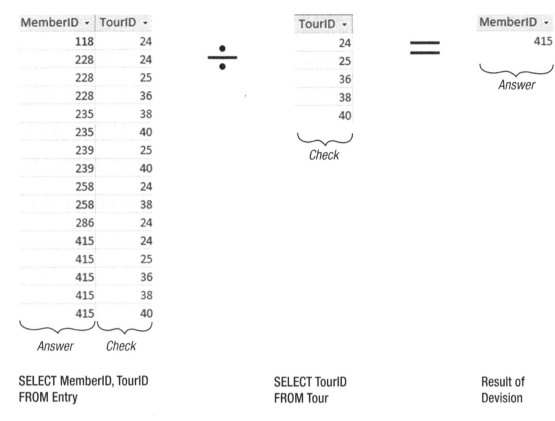

Figure 7-25. *Using division to find members who have entered all tournaments*

Looking at Figure 7-25, we have in the middle a table with all the TourID values (I've labelled that *Check*). The division operation checks the left-hand table to find the values of MemberID that have a row for every TourID. The *Answer* (on the right of the figure) contains the MemberID values for members who have entered every tournament. Member 415 can be found paired with each of the five tournaments in the Entry table, and so appears in the result of the division. Member 228 does not appear in the result because there are no rows in the Entry table with 228 paired with 38 or 40.

It is important to get the correct columns in the two tables involved in the division. I like to think of setting up the division operation like this:

- Decide which attribute I want to find out about. Let's call this *Answer*. In this case, I want to find values of MemberID, so our *Answer* attribute is MemberID.

- On the right-hand side of the division operator, the attribute(s) in the table should be the thing I want to check against. Let's call this attribute(s) *Check*. In this case, the *Check* attribute is TourID. We can get all the values for TourID from the Tournament table.

- On the left-hand side of the division, I want a table containing the just the two sets of attributes *Answer* and *Check*, as shown in Figure 7-25. We need MemberID and TourID (in this case, which members have entered which tournament, and these come from the Entry table). It is important that these are the only two columns in the left-hand table. If extra columns are added, then we will be asking different questions, as explained in the next section.

As a small aside, many people wonder why this operation is called *division*, as it doesn't seem to relate particularly well to something like 4 divided by 2. Division is the inverse (or undoing) of multiplication in normal arithmetic. For set operations, division is like the inverse of the Cartesian product. If you think of taking the Cartesian product of the two tables in the middle and far right of Figure 7-25, you will get a table with the same columns (but not rows) as on the far left of Figure 7-25.

We can answer a number of questions by changing what is on the right-hand side of the division operator. For example, if we wanted to know who had entered all the Open tournaments, we would replace the table in the middle of Figure 7-25 with just the rows for Open tournaments:

```
SELECT TourID
FROM Tour
WHERE TourType = 'Open';
```

Projecting Appropriate Columns

As with intersection and difference operations, projecting different columns in division operations will give you answers to different questions. Once again, an example is the easiest way to understand this. In Figure 7-26, an extra column has been retrieved from the Entry table. Can you understand what this query is finding?

MemberID ↤	Year ↦	TourID ↤
118	2014	24
228	2015	24
228	2015	25
228	2015	36
235	2013	38
235	2015	38
235	2014	40
235	2015	40
239	2015	25
239	2013	40
258	2014	24
258	2014	38
286	2013	24
286	2014	24
286	2015	24
415	2015	24
415	2013	25
415	2014	36
415	2015	36
415	2013	38
415	2015	38
415	2013	40
415	2014	40
415	2015	40
415	2020	40

Answer   Check

÷

TourID ↦
24
25
36
38
40

Check

= ?

Answer

Figure 7-26. *What is the division operation finding?*

The division is looking for a set of *Answer* attributes in the left-hand table that are paired with every attribute from the Check table. In this case, the operation looks for a pair MemberID and Year in the left-hand table that appears with each of the tournaments. This division example is finding those members who have entered all tournaments in the same year.

SQL for Division

Using an output approach, the query we want can be expressed something like this:

Write out the value of m.LastName, m.FirstName from rows m in the Member table where for every row t in the Tournament table there exists a row e in the Entry table with e.MemberID = m.MemberID and e.TourID = t.TourID.

125

We have an SQL keyword for *exists* but not for *every*. We can get rid of the *every* word in the preceding statement by using the following slightly mind-bending logic. The phrase

*for **every row** t in the Tournament table **there exists a row** e in the Entry table...*

is equivalent to saying

***there is no row** t in the Tournament table **where there does not exist a row** e in the Entry table...*

Appendix 2 provides a more formal explanation of how to derive these expressions, but for now we will just rewrite the original description of how to retrieve the names of members who have entered all the tournaments by using the equivalence just discussed.

*Write out the value of m.LastName, m.FirstName from rows m in the Member table where ~~for every row~~ **there is no row** t in the Tournament table ~~there exists a row~~ **where there does not exist a row** e in the Entry table with e.MemberID = m.MemberID and e.TourID = t.TourID.*

The corresponding SQL is:

```
SELECT m.LastName, m.FirstName FROM Member m
WHERE NOT EXISTS
    (
     SELECT * FROM Tournament t
     WHERE NOT EXISTS
          (
           SELECT * FROM Entry e
           WHERE e.MemberID = m.MemberID AND e.TourID = t.TourID
          )
    );
```

The double negatives can be a bit daunting, but as I said at the beginning of the chapter, I promise a conceptually easier method to find members who have entered every tournament in the next chapter.

Summary

Because tables in a relational database have unique rows (if they are properly keyed!), they can be treated like mathematical sets. This allows us to use the set operations union, intersection, difference, and division.

Union, intersection, and difference are operations that act between union-compatible tables. This means the table on each side of the operator must have the same number of columns, and the columns must have the same domains (commonly interpreted as the same types). You can get union-compatible tables by sensibly projecting columns.

SQL has keywords to represent union, intersection, and difference, although not every implementation supports the keywords for all of these operations. If your SQL product does not support keywords for intersection or difference, you can find other ways to express the query. You should formulate your queries in the way you find most natural. Where you have very large amounts of data and speed is important, you may need to investigate the efficiencies of the different ways of formulating some queries.

Here is a summary of the set operations and alternative ways to represent them with SQL. A and B are two union-compatible tables with (for simplicity) just one column called `attribute`.

Union

A union operation returns all the unique rows that are in either table A or table B:

```
SELECT attribute FROM A
UNION
SELECT attribute FROM B;
```

Intersection

An intersection operation returns all rows that are in both table A and table B:

```
SELECT attribute FROM A
INTERSECT
SELECT attribute FROM B;
```

An alternative way to represent intersection is:

```
SELECT A.attribute
FROM A
WHERE EXISTS
    (SELECT B.attribute FROM B
     WHERE A.attribute = B.attribute);
```

Difference

Difference returns all rows that are in the first table (A) that are not in the second table (B). Some implementations use the keyword `MINUS` instead of `EXCEPT`:

```
SELECT attribute FROM A
EXCEPT
SELECT attribute FROM B;
```

An alternative way to represent difference is:

```
SELECT A.attribute
FROM A
WHERE NOT EXISTS
    (SELECT B.attribute FROM B
    WHERE A.attribute = B.attribute);
```

Division

The division operation helps with queries with the words *every* or *all*. Current versions of SQL do not support division directly. Refer to the sections "Division" and "Universal Quantifier and SQL" in Appendix 2 for details of how to express queries involving division.

For completeness, we repeat the following query, which returns the MemberID values for those members who have entered every tournament:

```
SELECT m.LastName, m.FirstName FROM Member m
WHERE NOT EXISTS
     (
      SELECT * FROM Tournament t
      WHERE NOT EXISTS
             (
              SELECT * FROM Entry e
              WHERE e.MemberID = m.MemberID AND e.TourID = t.TourID
             )
     );
```

CHAPTER 8

Aggregate Operations

SQL has a number of functions for counting, summing, averaging, and otherwise performing aggregate operations on a table. These functions enable us to perform a variety of queries. For example, we can count the number of members in the club or find the average handicap. We can group the data in different ways to find aggregates. For example, we might want to count the number of tournament entries in each individual year, or we might want to find the number of entries in each particular tournament.

In this chapter we will look at simple aggregates and how to make the most of the SQL grouping capabilities. In the next chapter we will look at window functions, which provide elegant solutions in situations that can be difficult to address with just the basic aggregate functionality.

Simple Aggregate Functions

Simple aggregates include averages, totals, and counts. These are straightforward ideas, but, as always, you need to be sure you understand how they work when nulls and duplicates are involved.

The COUNT() Function

The COUNT() function calculates the number of rows being returned from a query. The simplest example is to count all the rows returned by a query, which we can do by adding an asterisk between the parentheses. The following query will return the number of rows in the Member table:

```
SELECT COUNT(*)
FROM Member;
```

A single aggregate function such as COUNT() in the preceding query will return a table with one column and one row, as shown in Figure 8-1.

Expr1000 ▾
20

Figure 8-1. Result of COUNT() function

The output in Figure 8-1 was produced in Access, and, as you can see, the column is labelled with a default name. We can provide a better name by giving the column an alias. In the following query we have added a WHERE clause to count the subset of rows satisfying the condition Gender = 'F' and used an AS clause so the column has a more informative heading:

```
SELECT COUNT(*) AS NumberWomen
FROM Member
WHERE Gender = 'F';
```

The output of this query is shown in Figure 8-2.

Figure 8-2. *Result of COUNT() function with an alias*

Providing the columns returned by aggregates with an alias is a good idea so that the reader has some idea of what the numbers mean.

Managing Nulls

The previous query returns a count of the number of members with a gender of 'F'. Now consider the query here:

```
SELECT COUNT(*)
FROM Member
WHERE Gender <> 'F';
```

At first glance we might think that the two counts from the previous two queries should add up to the total number of members. But we need to be careful. In Chapter 2, we looked at how WHERE conditions operate when we make comparisons with a null (or empty) value. If there is no value for the attribute then we cannot say whether it does or does not satisfy a condition. We don't know! In SQL, if the value we are comparing is a null, then the result of the comparison will always be false. Rows in the table with a null in the Gender column will not be included in either of the two previous queries.

We could argue that the attribute should have been declared as NOT NULL in the design of the table. In Chapter 2 we discussed why this might not be a good idea. If we are trying to enter details for a new member who has not provided their gender then we either will be prevented from saving the details we do know or will have to guess the gender. Neither option is satisfactory. It's better to save the details and follow up on the missing data later.

We can explicitly find how many of the rows do not have a value for Gender with the query:

```
SELECT COUNT(*)
FROM Member
WHERE Gender IS NULL;
```

The numbers for the three previous queries with conditions Gender = 'F', Gender <> 'F', and Gender IS NULL will now add up to the total number of members in the club. Queries like the preceding one can be very useful for checking if there are null values in columns where we would ideally expect to have values.

The COUNT() function can also return the number of values in a particular column of a table or query. Let's look at a few of the columns in the Member table, as shown in Figure 8-3.

MemberID ▾	LastName ▾	FirstName ▾	Gender ▾	Handicap ▾	Coach ▾
118	McKenzie	Melissa	F	30	153
138	Stone	Michael	M	30	
153	Nolan	Brenda	F	11	
176	Branch	Helen	F		
178	Beck	Sarah	F		
228	Burton	Sandra	F	26	153
235	Cooper	William	M	14	153
239	Spence	Thomas	M	10	
258	Olson	Barbara	F	16	
286	Pollard	Robert	M	19	235
290	Sexton	Thomas	M	26	235
323	Wilcox	Daniel	M	3	
331	Schmidt	Thomas	M	25	153
332	Bridges	Deborah	F	12	235
339	Young	Betty	F	21	
414	Gilmore	Jane	F	5	153
415	Taylor	William	M	7	235
461	Reed	Robert	M	3	235
469	Willis	Carolyn	F	29	
487	Kent	Susan	F		

Figure 8-3. Some columns of the Member table

Say we want to find the number of members who have a coach. We have two options. One way is to formulate a query to return just those members who have a non-null value for Coach and count those:

```
SELECT COUNT(*)
FROM Member
WHERE Coach IS NOT NULL;
```

The other option is to ask the COUNT() function to specifically count the number of not-null values in the Coach column using COUNT(Coach):

```
SELECT COUNT(Coach)
FROM Member;
```

To recap: if we just want to find the number of rows returned from a query (or a whole table), we use SELECT COUNT(*). If we want to find the number of rows that have a value in a particular column, use SELECT COUNT(<*Column_Name*>). The COUNT(*) and COUNT(<*Column_Name*>) options allow us to be specific about how we want null values to be treated.

Managing Duplicates

The values in the Coach column of the Member table (Figure 8-3) are duplicated. There are only two distinct values (153 and 235). We therefore have two quite different questions that can be answered by counting the values in the Coach column: "How many people have a coach?" and "How many coaches are there?" The answer to the first question requires us to include all the values. The answer to the second question requires us just to count the distinct values. This can be done by including the DISTINCT keyword as in the query here:

```
--Won't work in Access (2016)
SELECT COUNT(DISTINCT Coach)
FROM Member;
```

While I am trying not to become product-specific in this book, I feel obliged (given how many copies of Access are in the world) to point out that Access does not currently support COUNT(DISTINCT). However, you can get the equivalent result in Access with the nested query that follows. (Note that SQL Server does not allow a subquery in the FROM clause for an aggregate.)

```
--Won't work in SQL Server (2012)
SELECT COUNT(*)
FROM (SELECT DISTINCT Coach FROM Member WHERE Coach IS NOT NULL);
```

You can also use the keyword ALL. This just reinforces that you want to count all values, rather than just distinct values. If you do not include either DISTINCT or ALL then all values are included by default.

Similar sorts of queries can be applied to the other tables in the golf club database. For example, we might want to know how many tournaments members entered in 2015 (11) or how many different tournaments members entered in 2015 (5). The two queries that follow will provide the answers to the respective questions:

```
-- How many tournaments were entered
SELECT COUNT(TourID)
FROM Entry
WHERE Year = 2015;
```

```
-- How many different tournaments were entered
SELECT COUNT(DISTINCT TourID)
FROM Entry
WHERE Year = 2015;
```

The AVG() Function

To find averages, we use the function AVG(). The parameter that goes in the parentheses (...), is the expression we want to average. The expression has to be a numeric value. If you try to average a text field such as LastName you will get an error.

As an example, we can find the average handicap for members of our club by including the Handicap column as the parameter for the AVG() function:

```
SELECT AVG(Handicap)
FROM Member;
```

The expression could be just the name of one of the numeric-valued columns as in the preceding query or it could be the result of a calculation. Say in another database we have an Order table that includes the columns Price and Quantity for each item ordered. The net value of each order can be found by multiplying the Price and Quantity. If we want to find the average net value for all our orders, we can put the expression Price * Quantity in the parentheses as seen here:

```
SELECT AVG(Price * Quantity)
FROM Order;
```

Managing Nulls

As with the COUNT() function, the AVG() function does not include rows where the value of the expression is null. In the Member table we have 20 members in total, and 17 members with handicaps. If we sum all the handicaps, we get 287. The AVG() function will take the total of the handicaps (287) and divide by the number of rows that have a non-null value in the Handicap column (17). This is what we want. If we included the members without handicaps (by dividing by the total number of rows, 20), we would essentially be saying that these members have a handicap of 0 by default. This would seriously skew the results.

It is not always so obvious whether you want the null values considered. For example, say we have another database with a table called Student and a column called TestScore. If we enter test scores for students, and some of the students do not take the test, then we will have a null in the TestScore column for those students. What do we really want for the average? We could take the average over all the students (divide the total score by the count of all students), which means the students who missed the test are effectively being counted as having scored 0. On the other hand, we might take the average of just those who participated in the test (divide by the number who took the test). AVG(TestScore) will always give us just the average for those who took the test. It is by no means trivial to determine which of the two options you need. There is a constant debate in schools as to whether the pass rates (on which funding may depend) should include students who have dropped out along the way.

If we want the average over all the students, including those with a null mark (counted as 0), we can calculate that in the query by totaling the marks (using the SUM() function) and dividing by the total number of students as seen here:

```
SELECT SUM(TestMark)/COUNT(*)
FROM Student;
```

We could have entered a mark of 0 for those students who did not take the test, saving us this complication. However, if we do that then we can no longer distinguish students who took the test and got 0 from students who missed the test. Regardless of whether that is an issue or not, it is always useful to be aware of the implications.

Managing Duplicates

As with the COUNT() function, the AVG() function can also incorporate the keywords ALL and DISTINCT. Just be aware that ALL (which is the default) means all the not-null values including duplicates, as opposed to only the distinct not-null values. It doesn't mean take an average over all the rows (including those that are null), as in our discussion in the previous section. I find it quite difficult to come up with examples of when you would want to average over just the distinct values—certainly none that apply to our club database.

Managing Types and Output

The AVG() function will accept only numeric expressions as a parameter. We cannot successfully average FirstName or JoinDate (although we could use functions to average the length of members' first names or the number of days since their join date).

What result do we expect to get when we average the handicaps of our members? The total of the handicaps is 287, and the number of people with handicaps is 17. The result for the AVG(Handicap) function in SQL Server 2012 is 16. The result in Access 2016 is 16.8823529411765. Why?

In SQL Server (and some other implementations of SQL), the average function returns the same type as the numbers being averaged. In this case, the Handicap column is an integer type, and so AVG(Handicap) in SQL Server returns an integer. It also does an integer division (which means the result is truncated to 16 rather than rounded up to 17). In Access the average is returning a floating-point number (i.e., one with a fractional part).

We can control how the result is calculated. If we want a result with a fractional part for the average, we can convert the Handicap value to a floating-point number before we do the average. To do this we can use the CONVERT() function that we mentioned in Chapter 7:[1]

```
SELECT AVG(CONVERT(FLOAT,Handicap))
FROM Member;
```

Another way to do this is just to multiply the handicap by 1.0, which effectively converts it to a floating-point:

```
SELECT AVG(Handicap * 1.0)
FROM Member;
```

The ROUND() Function

While not strictly speaking an aggregate function, it is worthwhile to take a moment to look at how to perform rounding. Because averaging involves a division by the number of items involved, the AVG() function will often return a result with many decimal places. We use a rounding function to specify the number of decimal places we would like included in the output of AVG() and other expressions that result in floating-point numbers. We provide the ROUND() function with two parameters: the expression to be rounded and the number of decimal places to return. The following statement returns the average handicap rounded to two decimal places:

```
SELECT ROUND(AVG(Handicap * 1.0), 2)
FROM Member;
```

Rounding can behave differently in different implementations of SQL. In Access the previous query will return 16.88, whereas in SQL Server it will return 16.880000. While it is possible to remove the trailing zeroes in SQL Server, it is often better to leave that sort of formatting to front-end tools such as report writers.

There are many different ways to carry out rounding, so it is a good idea to consult the documentation to understand how your implementation of SQL goes about it. The traditional method of everything ending in a 5 or greater to be rounded up (e.g., 4.5 rounds up to 5) causes a bias to higher numbers. To remove this bias some implementations of rounding functions round to the nearest *even* number. For example, 3.5 and 4.5 would both round to 4. This evens things out, but it can come as a surprise if you are not expecting it. SQL Server's ROUND() function rounds all the 5s up, while Access rounds the 5s to the nearest even number.

---

[1]Different versions of SQL will have different functions to do this. In Oracle, you might consider using the CAST function.

Other Aggregate Functions

SQL also provides other common aggregate functions such as SUM(), MAX(), and MIN(), which are very straightforward to use. Similar to the AVG() function, the arguments to the SUM() function must be a numeric expression (either a numeric attribute or some expression with a numeric result, such as Price * Quantity). The arguments to MAX() and MIN() can be numeric, text, or date types. For text types, the order is alphabetical. For dates, the order is chronological. For example, MIN(LastName) would return the first value of LastName alphabetically, while MAX(JoinDate) would return the most recent value of JoinDate.

It is possible to combine several aggregate functions in one query. The following query returns the maximum, minimum, and average values for Handicap.

```
SELECT MAX(Handicap) AS maximum, MIN(Handicap) AS minimum,
    ROUND(AVG(Handicap * 1.0),2) AS average
FROM Member;
```

Providing an alias for the result of each column with an AS clause helps make the result easier to understand. Figure 8-4 shows some typical output.

maximum ▾	minimum ▾	average ▾
30	3	16.88

Figure 8-4. *Typical output from a query with several aggregate functions*

Grouping

If we want to know how many times a particular member has entered tournaments we can query the Entry table. For example, if we would like to find how many times member 235 has entered tournaments, we could select all the rows in the Entry table for that member and count them as in the following query:

```
SELECT COUNT(*) AS NumEntries
FROM Entry
WHERE MemberID = 235;
```

If we want to find the number of entries for a different member, we would need to rewrite the query with a different WHERE clause. If we want to find the counts for all members, that would get very tedious.

Grouping allows us to find the counts for all members using one SQL statement. The key phrase GROUP BY is used to do this. Have a look at the following query:

```
SELECT COUNT(*) AS NumEntries
FROM Entry
GROUP BY MemberID;
```

The extra GROUP BY clause says, "Rather than just count all the rows in the Entry table, count all the subsets or groups with the same MemberID." Figure 8-5 illustrates how we can visualize what is happening.

We can also include the fields we are grouping by in the SELECT clause so we can see which counts belong to which entries, as in the query here:

```
SELECT MemberID, COUNT(*) AS NumEntries
FROM Entry
GROUP BY MemberID;
```

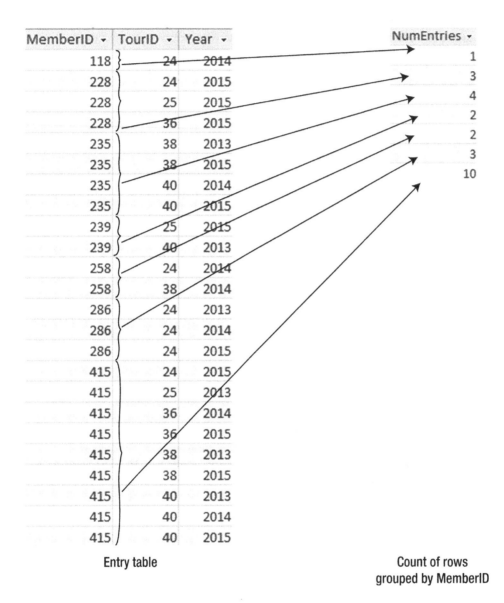

Figure 8-5. *Counting the rows in the Entry table grouped by MemberID*

The output of this query is shown in Figure 8-6.

MemberID ▾	NumEntries ▾
118	1
228	3
235	4
239	2
258	2
286	3
415	10

Figure 8-6. *Including the MemberID in the output*

We might prefer to see the names of the members in the output in Figure 8-6. In this case, we need to join the Entry table with the Member table first, and then group and count.

```
SELECT m.MemberID, m.LastName, m.FirstName, COUNT(*) AS NumEntries
FROM Entry e INNER JOIN Member m ON m.MemberID = e.MemberID
GROUP BY m.MemberID, m.LastName, m.FirstName;
```

The output is shown in Figure 8-7.

MemberID ▾	LastName ▾	FirstName ▾	NumEntries ▾
118	McKenzie	Melissa	1
228	Burton	Sandra	3
235	Cooper	William	4
239	Spence	Thomas	2
258	Olson	Barbara	2
286	Pollard	Robert	3
415	Taylor	William	10

Figure 8-7. *Joining the Entry and Member tables and grouping by the IDs and names*

You might wonder why we have included LastName and FirstName in the GROUP BY clause in the preceding query. When you are using GROUP BY, the SELECT clause can only include the fields you are grouping by or the aggregates. If we want to see the names in the output, we need to include them in the fields we are grouping by. This guards against cases where there might be different names for one MemberID (clearly impossible in this case, as MemberID is the primary key of the Member table). Putting this aside for now, if there were two rows with a MemberID of 118 with two different names, then if we group just by MemberID, it would not be possible to determine which name to associate with the count in the output in Figure 8-7.

We can get a range of different information from our data using GROUP BY if we include WHERE clauses and different attributes in the GROUP BY clause. Let's take another look at the Entry table. If we want find the number of entries for each tournament, we imagine grouping all the rows with the same TourID together and then counting the rows in each set, as in the query here:

```
SELECT TourID, COUNT(*) AS NumEntries
FROM Entry
GROUP BY TourID;
```

The output is shown in Figure 8-8.

TourID ▾	NumEntries ▾
24	7
25	3
36	3
38	5
40	7

Figure 8-8. *Counting the number of entries in each tournament*

We do not always need to count all the rows in the table. We might like to select a subset of the rows first. For example, we might just want to gather the statistics in Figure 8-8 just for the year 2014. The following query shows the SQL to do this. Note that the WHERE clause (which finds the subset of the rows we want to consider) must come before the GROUP BY clause:

```
SELECT TourID, COUNT(*) AS NumEntries
FROM Entry
WHERE Year = 2014
GROUP BY TourID;
```

By adding more fields in the GROUP BY clause, we can get more detailed information. If we want to repeat this query for each tournament for each year, we can remove the WHERE clause and group by both Year and TourID:

```
SELECT TourID, Year, COUNT(*) AS NumEntries
FROM Entry
GROUP BY TourID, Year;
```

Figure 8-9 shows how the grouping on both fields works.

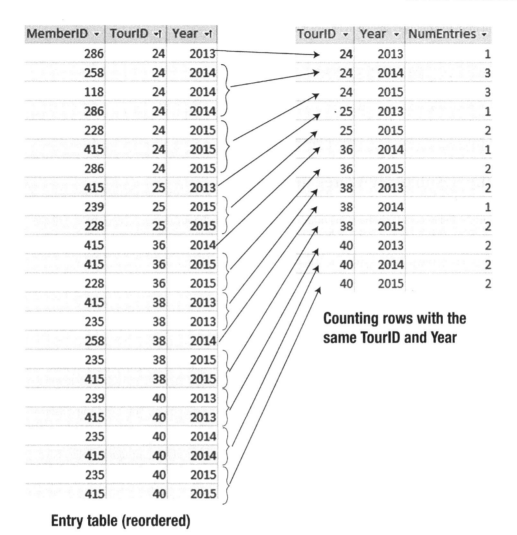

Entry table (reordered)

Figure 8-9. *Grouping by TourID and Year*

We can use grouping with aggregate functions other than COUNT(). For example, if we want to see the maximum, minimum, and average handicap for women and men, we could use a query like the one here:

```
SELECT Gender, MIN(Handicap)as Minimum, Max(Handicap)as Maximum,
       ROUND(AVG(Handicap),1) AS Average
FROM Member
GROUP BY Gender;
```

The output from this query is shown in Figure 8-10.

Gender ▾	Minimum ▾	Maximum ▾	Average ▾
F	5	30	18.8
M	3	30	15.2

Figure 8-10. *Grouping aggregates for Handicap by Gender*

Filtering the Result of an Aggregate Query

Once we have calculated some aggregates for groups of rows, we may want to ask some questions about the results. For example, in Figure 8-9, we have the number of entries in each tournament in each year. A likely question is "Which tournaments had three or more entries?" Looking at the result table in Figure 8-9, we want to select just those rows in the aggregated output with a count greater than or equal to 3. We can do this with the HAVING keyword. Take a look at the following query:

```
SELECT TourID, Year, COUNT(*) AS NumEntries
FROM Entry
GROUP BY TourID, Year
HAVING COUNT(*) >= 3;
```

The HAVING clause always comes after a GROUP BY clause. The aggregate and the grouping is carried out first, and the output rows matching some condition (in this case, COUNT(*) >= 3) are selected. It is like having a WHERE clause that acts on the aggregated numbers. As a little aside, we must use COUNT(*) in the HAVING clause; we can't use the alias NumEntries from the first line of the statement. This alias is just used at the end of the query to label the output column.

Let's look at another example. Say we want to find those members who have entered four or more tournaments. First, construct a set of rows with the members and the counts of the tournaments they have entered, as in the first three lines of the query that follows. We then use the HAVING clause to select just those rows from the result with COUNT(*) >= 4:

```
SELECT MemberID, COUNT(*) AS NumEntries
FROM Entry
GROUP BY MemberID
HAVING COUNT(*) >= 4;
```

We have two opportunities to select a subset of rows in queries involving aggregates. If we take the subset *before* we do the aggregation, we use a WHERE clause. When we want to select just some rows *after* the aggregation, we use a HAVING clause. For example, let's change the previous query to find out which members have entered more than four Open tournaments. To find the Open tournaments, we need to do the following:

1. Join the Entry table with the Tournament table.

2. Take just the subset of entries for Open tournaments (with a WHERE clause).

3. Group the entries for each member and count them.

4. Take the resulting aggregate table and retrieve just those rows with a count greater than 4 (with a HAVING clause).

The process is illustrated in Figure 8-11.

MemberID	e.TourID	Year	t.TourID	TourType
~~118~~	~~24~~	~~2014~~	~~24~~	Social
~~228~~	~~24~~	~~2015~~	~~24~~	Social
~~258~~	~~24~~	~~2014~~	~~24~~	Social
~~286~~	~~24~~	~~2013~~	~~24~~	Social
~~286~~	~~24~~	~~2014~~	~~24~~	Social
~~286~~	~~24~~	~~2015~~	~~24~~	Social
~~415~~	~~24~~	~~2015~~	~~24~~	Social
~~228~~	~~25~~	~~2015~~	~~25~~	Social
~~239~~	~~25~~	~~2015~~	~~25~~	Social
~~415~~	~~25~~	~~2013~~	~~25~~	Social
228	36	2015	36	Open
415	36	2014	36	Open
415	36	2015	36	Open
235	38	2013	38	Open
235	38	2015	38	Open
258	38	2014	38	Open
415	38	2013	38	Open
415	38	2015	38	Open
235	40	2014	40	Open
235	40	2015	40	Open
239	40	2013	40	Open
415	40	2013	40	Open
415	40	2014	40	Open
415	40	2015	40	Open

**1. Join Entry and Tournament
Select rows for Open
tournaments
(WHERE)**

MemberID	e.TourID	Year	t.TourID	TourType
228	36	2015	36	Open
235	40	2015	40	Open
235	40	2014	40	Open
235	38	2015	38	Open
235	38	2013	38	Open
239	40	2013	40	Open
258	38	2014	38	Open
415	40	2015	40	Open
415	40	2014	40	Open
415	40	2013	40	Open
415	38	2015	38	Open
415	38	2013	38	Open
415	36	2015	36	Open
415	36	2014	36	Open

**2. Group by MemberID
and count**

MemberID	NumEntries
~~228~~	~~1~~
235	4
~~239~~	~~1~~
~~258~~	~~1~~
415	7

**3. Retain rows from result
with COUNT >= 4
(HAVING)**

Figure 8-11. *Finding members who have entered more than four Open tournaments*

The query for the process in Figure 8-8 is:

```
SELECT MemberID, COUNT(*) AS NumEntries
FROM Entry e INNER JOIN Tournament t ON e.TourID = t.TourID
WHERE t.TourType = 'Open'
GROUP BY MemberID
HAVING COUNT(*) > 4;
```

In Chapter 2 we looked at ordering the output of a query. We can also order the output by the aggregate. If we would like to see results of the previous query in descending order of the number of tournaments entered, we could add an ORDER BY COUNT(*) DESC clause at the end of the query.

Using Aggregates to Perform Division Operations

In Chapter 7, we looked at the algebra operation division. To recap, division allows us to answer many questions containing the words *all* or *every*. For example, say we want to find those members who have entered *every* tournament. Figure 8-12 reviews how we can use division to do this. The attribute we want returned in our *Answer* is MemberID. On the right side of the division operator, we have the set of things to *Check* against (in this case, a list of all the TourID values projected from the Tournament table). On the left side of the division operator is a table that has both the attributes from *Answer* and *Check* (in this case, the columns MemberID and TourID from the Entry table). The result of the division is a list of the MemberID values that appear with every tournament (in this case, just the one member with ID 415).

This figure is the same as Figure 7-25.

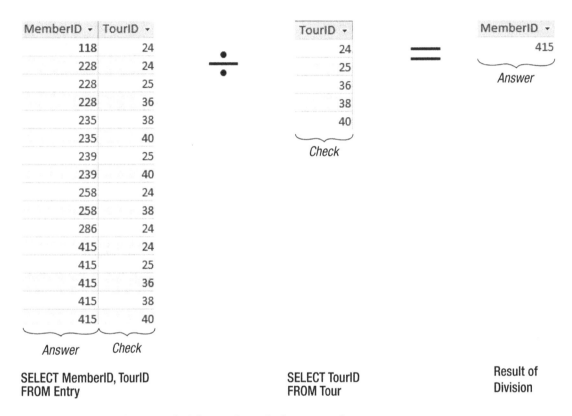

Figure 8-12. *Using division to find the members who have entered every tournament*

Currently, standard implementations of SQL do not have a keyword for the division operation, so we need to find other ways to express a query like that depicted in Figure 8-12. We looked at one way in Chapter 7 and some others in Appendix 2. Here, we will look at a method that uses aggregates.

The Tournament table lists five different tournaments. If we can find a member who has entered five *different* tournaments, then he or she must have entered all of them. We now have the ability to use aggregates and grouping to construct the equivalent of a division operation.

We have already seen queries to count how many tournaments each member has entered. However, now we want to count only the *different* tournaments entered by each member. Adding the DISTINCT keyword in the COUNT() function will achieve this:

```
SELECT MemberID, COUNT(DISTINCT TourID) AS NumTours
FROM Entry e
GROUP BY MemberID;
```

The result of this query is shown in Figure 8-13.

MemberID ▾	NumTours ▾
118	1
228	3
235	2
239	2
286	2
415	5

Figure 8-13. *Finding the number of distinct tournaments entered by each member*

From the resulting table in Figure 8-13, we now want just those rows where the NumTours is equal to the number of distinct tournaments, which is 5 in this case. We can use the HAVING clause to find those members who have entered five different tournaments:

```
SELECT MemberID
FROM Entry e
GROUP BY MemberID
HAVING COUNT(DISTINCT TourID) = 5;
```

We can make this query more general by replacing 5 with an expression to calculate the number of distinct tournaments on the fly:

```
SELECT MemberID
FROM Entry e
GROUP BY MemberID
HAVING COUNT(DISTINCT TourID) =
    (SELECT COUNT(DISTINCT TourID) FROM Tournament);
```

This query is equivalent to the algebra division operation as depicted in Figure 8-12. It returns the IDs of members who have entered every tournament. To summarize, we count the number of distinct tournaments each member has entered, and then, using the HAVING clause, retain just those whose count equals the number of possible tournaments (a distinct count from the Tournament table). I find this method of doing a division conceptually more straightforward than the ones in Chapter 7 and Appendix 2. However, all methods accomplish the same goal.

Nested Queries and Aggregates

We have already lightly covered nested queries and aggregates in Chapter 4. It is useful to revisit this idea here. In this chapter, we've looked at how to find averages, totals, counts, and so on. Now we can use these aggregate results in other queries. For example, we might want to find everyone with a handicap greater than the average handicap. Consider the following query:

```
SELECT * FROM Member
WHERE Handicap >
      (SELECT AVG (Handicap)
       FROM Member);
```

The inner part of the query returns the average handicap, and the outer part of the query compares the handicap of each member with that average.

Let's try something else. What about finding members who have entered more than three tournaments? If your mind goes blank, you can revert to the outcome approach of picturing the tables and figuring out what the rows you want returned will look like. Figure 8-14 shows how you can visualize the query.

MemberID ▾	LastName ▾	FirstName ▾	Gender ▾
118	McKenzie	Melissa	F
138	Stone	Michael	M
153	Nolan	Brenda	F
176	Branch	Helen	F
178	Beck	Sarah	F
228	Burton	Sandra	F
235	Cooper	William	M
239	Spence	Thomas	M
258	Olson	Barbara	F
286	Pollard	Robert	M

MemberID ▾	TourID ▾	Year ▾
118	24	2014
228	24	2015
228	25	2015
228	36	2015
235	38	2013
235	38	2015
235	40	2014
235	40	2015
239	25	2015
239	40	2013
258	24	2014

Figure 8-14. *Which members have more than three entries in tournaments?*

We can describe the members we want returned like this:

Find all the rows m from the Member table where if we count the number of rows e from the Entry table for that member (m.MemberID = e.MemberID) the count is > 3.

This turns into SQL in a straightforward way, as shown here:

```
SELECT * FROM Member m
WHERE
      (SELECT COUNT (*)
       FROM Entry e
       WHERE e.MemberID = m.MemberID) > 3;
```

What about something a bit more complex? How do we find the average number of tournaments entered by members? We will need the AVG() function, but what are we trying to average? We want to count the number of tournaments for each member and then average those counts.

We can use grouping, as described in the previous section, to find the number of tournaments entered by each member:

```
SELECT MemberID, COUNT (*) AS CountEntries FROM Entry
GROUP BY MemberID
```

The result of the preceding query is shown in Figure 8-15.

MemberID ▾	CountEntries ▾
118	1
228	3
235	4
239	2
258	2
286	3
415	9

Figure 8-15. *Number of entries for each member*

Now we want to find the average of the column CountEntries. As a first try, it seems reasonable to use our previous query as the inner part of a nested query, and then attempt to find the average:

```
--Won't work in some implementations
SELECT AVG (CountEntries) FROM
      (SELECT MemberID, COUNT (*) AS CountEntries FROM Entry
       GROUP BY MemberID);
```

However, many versions of SQL do not support a nested query in a FROM clause. The preceding query works fine in Access 2013 but not in some other implementations of SQL.

We encountered a similar problem in the previous chapter when we wanted to join a table with the result of a union. We simply give the result of the inner query an alias (e.g., NewTable). This creates a temporary virtual table (often referred to as a *derived* table). We can then access the attributes of the new virtual table as here:

```
SELECT AVG (NewTable.CountEntries) FROM
      (SELECT MemberID, COUNT (*) AS CountEntries FROM Entry
       GROUP BY MemberID)AS NewTable;
```

Summary

Aggregate functions provide us with the means to answer a huge range of questions about our data. Here is a summary of some of the main points in this chapter.

Most versions of SQL will offer the simple aggregate functions MIN(), MAX(), COUNT(), SUM(), and AVG().

- For COUNT(), you often just want to count rows returned by a query. This can be done by including an asterisk in the parentheses: COUNT(*). If you include a column name in the parentheses (e.g., COUNT(Handicap)) then only the non-null values in that column will be included in the count.

- For the other common aggregates, you need to include a field name. For AVG() and SUM() this needs to be a numeric expression, such as AVG(Handicap).

Nulls and duplicates:

- Null values are not included when calculating aggregates. For example, AVG(Handicap) is the sum of the handicaps divided by the number of rows that have a non-null value for Handicap.

- By default, all non-null values are included in the aggregates. You can include the keyword DISTINCT to remove duplicates. For example, COUNT(DISTINCT Handicap) will count the number of different values appearing in the Handicap column.

Grouping:

- The GROUP BY clause can be used to collect rows with the same value of some expression together and then apply the aggregates to those groups. For example, we can find the number of tournaments each member has entered by grouping together all the rows in the Entry table with the same value of MemberID (e.g., SELECT MemberID, COUNT(*) FROM Entry GROUP BY MemberID).

- After you have grouped and performed an aggregate, you can select rows from the resulting table using the keyword HAVING. For example, we can find members who have entered three or more tournaments by adding the clause HAVING COUNT(*) >= 3 to the expression in the previous item.

- Use WHERE to select a subset of rows *before* the grouping and aggregating. Use HAVING to select a subset of rows *after* the grouping and aggregating.

More complex aggregates:

- Use derived tables where you want to nest aggregates, such as to find the average of counts. Simply give the inner query an alias.

- Compare counts of rows to do the equivalent of relational division.

CHAPTER 9

Window Functions

Window functions were added to standard SQL in 2003 and provide considerable extra capability for dealing with aggregation. Window functions allow us to perform aggregates on a "window" of data based upon the current row. This provides an elegant way to specify queries to perform actions such as ranking, running totals, and rolling averages. Window functions also provide considerable flexibility when it comes to grouping data for aggregation as they allow a single query to have several different groups or partitions. It is also possible to reference the data contributing to the aggregate from within the query. This allows the underlying data to be compared to the aggregate.

Oracle and Postgres have supported window functions for many years, while SQL Server just introduced them in 2012. Access and MySQL do not currently support these functions. This chapter outlines how to use a few of the most common window functions.

Simple Aggregates

To get started with window functions we will use them to write alternate queries for some of the simple aggregates we encountered in Chapter 8. Let's reconsider a simple aggregate query to count and average members' handicaps:

```
SELECT COUNT(Handicap) AS Count, AVG(Handicap * 1.0)as Average
FROM Member;
```

The output for the query is shown in Figure 9-1.

Count ▾	Average ▾
17	16.88

Figure 9-1. *Output for simple count and average of handicaps*

With simple aggregates the only attributes allowed in the SELECT clause are the aggregate and those attributes included in a GROUP BY clause. This means we no longer have access to the individual handicaps contributing to the results.

Window functions allow us to retrieve the underlying data along with the aggregates. The keyword for window functions is OVER(); they are also sometimes referred to as *over functions*.

Here is a query similar to the preceding one using the OVER() function:

```
SELECT MemberID, LastName, FirstName, Handicap,
    COUNT(Handicap) OVER() AS Count,
    AVG(Handicap * 1.0) OVER() as Average
FROM Member;
```

Unlike the simple COUNT() function, with the OVER() function we are able to include additional fields in the SELECT clause. In the preceding query we have included four fields of detailed data about each member along with the two aggregates (which are indented on new lines to make them easier to read). The aggregates are just the same as the simple aggregates but include the OVER() function.

Part of the output of the preceding query is shown in Figure 9-2. The count and average of the handicaps appear with the detailed data for each member.

MemberID ▾	FirstName ▾	LastName ▾	Handicap ▾	Average ▾	Count ▾
118	Melissa	McKenzie	30	16.88	17
138	Michael	Stone	30	16.88	17
153	Brenda	Nolan	11	16.88	17
176	Helen	Branch		16.88	17
178	Sarah	Beck		16.88	17
228	Sandra	Burton	26	16.88	17
235	William	Cooper	14	16.88	17
239	Thomas	Spence	10	16.88	17
258	Barbara	Olson	16	16.88	17
286	Robert	Pollard	19	16.88	17
290	Thomas	Sexton	26	16.88	17

Figure 9-2. *Output when using OVER() to count and average handicaps*

While it doesn't seem particularly useful in the example in Figure 9-2 to have the aggregates returned for every row, it opens the door to some new queries. We are now able to easily compare each individual's handicap with the average, something that was not at all simple without window functions. On the third line of the following query we subtract the average of the handicap from the handicap for each member and include that in the SELECT clause:

```
SELECT MemberID, LastName, FirstName, Handicap,
    AVG(Handicap * 1.0) OVER() AS Average,
    Handicap - AVG(Handicap *1.0) OVER() AS Difference
FROM Member;
```

The result is displayed in Figure 9-3.

MemberID ▾	FirstName ▾	LastName ▾	Handicap ▾	Average ▾	Difference ▾
118	McKenzie	Melissa	30	16.88	13.12
138	Stone	Michael	30	16.88	13.12
153	Nolan	Brenda	11	16.88	-5.88
176	Branch	Helen		16.88	
178	Beck	Sarah		16.88	
228	Burton	Sandra	26	16.88	9.12
235	Cooper	William	14	16.88	-2.88
239	Spence	Thomas	10	16.88	-6.88

Figure 9-3. *Window functions allow us to compare aggregates with detail values*

Partitions

The OVER() function can also be used to produce queries that are similar to the GROUP BY queries we looked at in the previous chapter. The key phrase we need here is PARTITION BY. Let's try some different counts on rows in the Entry table. If we use just the function OVER() with our COUNT(*) function, we will count *all* the rows, whereas if we use OVER(PARTITION BY TourID) it will count the rows for each different value of TourID.

The real power of partitioning is that, unlike the GROUP BY clause for simple aggregates, it is possible to have several different partitions in a single query. This is best explained by an example. The following query includes three different counts:

```
SELECT MemberID, TourID, Year,
COUNT(*) OVER() as CountAll,
COUNT(*) OVER(PARTITION BY TourID) AS CountTour,
COUNT(*) OVER(PARTITION BY TourID, Year) AS CountTourYear
FROM Entry;
```

The output is shown in Figure 9-4.

MemberID	TourID	Year	CountAll	CountTour	CountTourYear
286	24	2013	24	7	1
286	24	2014	24	7	3
118	24	2014	24	7	3
258	24	2014	24	7	3
228	24	2015	24	7	3
286	24	2015	24	7	3
415	24	2015	24	7	3
415	25	2013	24	3	1
228	25	2015	24	3	2
239	25	2015	24	3	2
415	36	2014	24	3	1
228	36	2015	24	3	2
415	36	2015	24	3	2
235	38	2013	24	5	2
415	38	2013	24	5	2
258	38	2014	24	5	1
415	38	2015	24	5	2
235	38	2015	24	5	2
415	40	2013	24	6	2
239	40	2013	24	6	2
415	40	2014	24	6	2
235	40	2014	24	6	2
415	40	2015	24	6	2
235	40	2015	24	6	2

Figure 9-4. *Using different partitions in a single query*

In Figure 9-4 the column CountAll displays the result of COUNT(*) OVER(), which counts every row in the Entry table (24).

The column CountTour is the result of COUNT(*) OVER(TourID), which partitions (or groups) the rows with the same value of TourID and then counts them. The top three sets of solid boxes in Figure 9-4 show the rows contributing to CountTour for TourID of 24, 25, and 36.

The column CountTourYear is the result of COUNT(*) OVER(TourID, Year) and partitions all the rows with the same values for TourID and Year. The set of dashed boxes toward the bottom of Figure 9-4 shows examples of how these counts are evaluated.

Order By Clause

The OVER() function can include an ORDER BY clause. This specifies an order for the rows to be visited when the aggregates are evaluated. Having an order for the rows provides a mechanism for carrying out running totals and ranking operations.

Cumulative Aggregates

If an ORDER BY clause is included in the OVER() function then, by default, the aggregate is carried out from the beginning of the partition to the current row (but see below for a more precise definition).

Have a look at the following query:

```
SELECT MemberID, TourID, Year,
COUNT(*) OVER(ORDER BY Year) AS Cumulative
FROM Entry;
```

In the Entry table we have several rows with the same value of Year (as you can see in Figure 9-5). As far as the ordering goes, these rows are equivalent, so if one of them is included in a count then we should include them all. I'll now correct the definition of what rows are included in the aggregate.

If an ORDER BY clause is included in the OVER() function then, by default, the aggregate is carried out from the beginning of the partition to the current row, and includes any following rows with the same value of the ordering expression.

The output in Figure 9-5 illustrates what this means.

MemberID ▾	TourID ▾	Year ▾	Cumulative ▾
286	24	2013	6
239	40	2013	6
415	25	2013	6
415	38	2013	6
415	40	2013	6
235	38	2013	6
118	24	2014	13
415	40	2014	13
415	36	2014	13
258	24	2014	13
258	38	2014	13
235	40	2014	13
286	24	2014	13
415	38	2015	24
228	24	2015	24
228	25	2015	24
228	36	2015	24
235	40	2015	24
239	25	2015	24
235	38	2015	24
286	24	2015	24
415	24	2015	24
415	36	2015	24
415	40	2015	24

Figure 9-5. *Using ORDER BY to produce a cumulative count for each year*

In Figure 9-5 the rows are ordered by Year. Let's see how this cumulative counting works for the first few rows. For the first row, if we count from the beginning of the table we have 1 row. However, the next five rows have the same value for our ordering expression Year, so we include them in the count, giving us a total of 6.

Now let's move down to the first row for member 258. Counting from the beginning of the table we have 10 rows, but the next 3 rows have the same value of Year. This makes a total of 13.

Essentially, we have a cumulative count of entries for each year. We have 6 entries in the first year (solid boxes), and for the second year we have an additional 7 entries to make 13 total (dashed boxes).

The SUM() function works in much the same way if there is an ORDER BY clause in the OVER() function to give us running totals.

Let's say the club collects data on income from fundraising and tournaments in a table called Income. Figure 9-6 shows income for the first six months.

Month ▾	Income ▾
1	2400
2	3800
3	1400
4	4500
5	6200
6	4800

Figure 9-6. *Income table*

We can find a running total of the income by performing a SUM(Income) with an ORDER BY Month clause in the OVER() function, as in the following query:

```
SELECT Month, Income,
SUM(Income) OVER(ORDER BY Month) AS RunningTotal
FROM Income;
```

The income is summed from the beginning of the table to the current row (when ordered by the value of Month), as shown in Figure 9-7.

Month ▾	Income ▾	RunningTotal ▾
1	2400	2400
2	3800	6200
3	1400	7600
4	4500	12100
5	6200	18300
6	4800	23100

Figure 9-7. *Running totals for monthly income ordered by month*

Ranking

Yet another use for the ORDER BY clause is with the RANK() function. As an example we will rank the members of the club by their handicap. Have a look at the following query:

```
SELECT MemberID, Handicap,
RANK() OVER (ORDER BY Handicap) AS Rank
FROM Member
WHERE Handicap IS NOT NULL;
```

The ORDER BY clause in the OVER() function specifies the order of the rows when determining the rank–in this case the value of Handicap. Each time the value of Handicap changes, the rank becomes the row number in the partition (in this case the entire table ordered by Handicap).The rank then stays the same until the value of Handicap changes, as shown in Figure 9-8. (Some of the handicaps have been changed to illustrate the process more clearly. Rows with the same value of Handicap and therefore rank have been delineated.)

MemberID ▾	Handicap ▾	Rank ▾↑
461	3	1
323	3	1
414	5	3
415	7	4
239	7	4
153	7	4
332	12	7
235	14	8
258	14	8
286	19	10
339	21	11
331	25	12
290	26	13
228	26	13
469	26	13
138	30	16
118	30	16

Figure 9-8. *Result of the RANK() function ordered by handicap. Rows with the same value handicap have the same rank*

The first row in Figure 9-8 has rank 1 (it is the first row!). The second row has the same value of the order expression (Handicap) as the previous row so it also has rank 1. In the next row the value of Handicap has changed so the rank becomes the row number (3).

Null values will be included in the ranking, which is why they have been explicitly excluded in the previous query. Without the WHERE clause the null values would have been included at the top of the ranking (or at the bottom if the order had been DESC).

Combining Ordering with Partitions

In the previous sections on ordering I didn't include any partitions in the queries. Now that we (hopefully) understand the concept we can look at some more examples.

Let's consider a more detailed Income table that has monthly amounts for each of three areas where the golf club carries out fundraising. The data for the first five months of the year is shown in Figure 9-9.

Month ▾	Area ▾	Income ▾
1	Halswell	2400
1	Pegasus	3868
1	Russley	2123
2	Halswell	3800
2	Pegasus	2719
2	Russley	3534
3	Halswell	1400
3	Pegasus	1650
3	Russley	1486
4	Halswell	4500
4	Pegasus	5072
4	Russley	4471
5	Halswell	6200
5	Pegasus	6406
5	Russley	5846

Figure 9-9. *Income table including areas*

We will build up some queries slowly.

First, we will just calculate the total income for the table. We could use a simple SUM() aggregate, but we will include an OVER() function so we can keep the detail in the output:

```
SELECT Month, Area, Income,
   SUM(Income) OVER() AS Total
FROM Income;
```

This will produce a table the same as in Figure 9-9 but with an additional column, Total, which will have the overall total for every row.

Now let's change this to a running total. We do this by including an ORDER BY clause in the OVER() function. By default this calculates the total for the values from the beginning of the table to the current row and the next rows with the same value of Month. The query is:

```
SELECT Month, Area, Income,
   SUM(Income) OVER(ORDER BY MONTH) AS RunningTotal
FROM Income;
```

The incomes are summed from the top of the table to the current row including the following rows with the same value of Month (the attribute we are ordering by). Essentially, the output sums all the values for each month and then accumulates the totals month by month. The output is shown in Figure 9-10. The different months have been delineated.

Month	Area	Income	RunningTotal
1	Halswell	2400	8391
1	Pegasus	3868	8391
1	Russley	2123	8391
2	Halswell	3800	21011
2	Pegasus	2719	21011
2	Russley	3534	21011
3	Halswell	1400	25547
3	Pegasus	1650	25547
3	Russley	1486	25547
4	Halswell	4500	39590
4	Pegasus	5072	39590
4	Russley	4471	39590
5	Halswell	6200	58042
5	Pegasus	6406	58042
5	Russley	5846	58042

Figure 9-10. *Running total when ordering by month*

Now let's look at the areas independently. This requires a PARTITION BY clause. Consider this query:

```
SELECT Month, Area, Income,
    SUM(Income) OVER(
        PARTITION By Area
        ORDER BY MONTH) AS AreaRunningTotal
FROM Income;
```

The PARTITION BY clause needs to come before the ORDER BY clause, which reflects what is happening. We first partition the data and then order within the partitions. The aggregate is calculated for rows from the beginning of the current partition to the current row. Figure 9-11 shows the output for just the first five months of the year. The three partitions have been delineated so it is easier to see what is happening.

Month ▾	Area ▾	Income ▾	AreaRunningTotal ▾
1	Halswell	2400	2400
2	Halswell	3800	6200
3	Halswell	1400	7600
4	Halswell	4500	12100
5	Halswell	6200	18300
1	Pegasus	3868	3868
2	Pegasus	2719	6587
3	Pegasus	1650	8237
4	Pegasus	5072	13309
5	Pegasus	6406	19715
1	Russley	2123	2123
2	Russley	3534	5657
3	Russley	1486	7143
4	Russley	4471	11614
5	Russley	5846	17460

Figure 9-11. *Running totals for income partitioned by area and ordered by month*

Framing

The last feature of the window functions we will look at is the ability to further specify which rows are included in an aggregate. This is how the name *window functions* came about. They provide a *window* or *frame* onto the section of data we are interested in. The general form of the OVER() function has three clauses, as shown here:

```
OVER(
    [ <PARTITION BY clause> ]
    [ <ORDER BY clause> ]
    [ <ROWS clause> ]
    );
```

We have already looked at two of these clauses: The PARTITION BY clause allows us to group the data by some expression before aggregating. The ORDER BY clause allows us to determine the order in which the aggregate function traverses the rows within a partition and allows us to perform ranking and running totals. A ROWS clause allows us to narrow down the set of rows, relative to the current row, that are to be included in the aggregate.

By default, a query with an OVER(ORDER BY) clause calculates the aggregate of the values from the beginning of the current partition up to and including the current row. Let's recap with a query that calculates a running average for each area:

```
SELECT Month, Area, Income,
AVG(Income) OVER (
    PARTITION BY AREA
    ORDER BY Month) AS AreaRunningAverage
FROM Income;
```

The output in Figure 9-12 is for just the Halswell area. The solid boxes show which rows are included in the average for the third row from the top. The dashed boxes show the rows contributing to the average for the third row from the bottom of the image. If there is no ROWS clause after an ORDER BY clause, then this is the default behaviour.

Month ▾	Area ▾	Income ▾	AreaRunningAverage ▾
1	Halswell	2400	2400
2	Halswell	3800	3100
3	Halswell	1400	2533
4	Halswell	4500	3025
5	Halswell	6200	3660
6	Halswell	4800	3850
7	Halswell	4300	3914
8	Halswell	4900	4038
9	Halswell	5200	4167
10	Halswell	4700	4220
11	Halswell	4500	4245
12	Halswell	5000	4308

Figure 9-12. Running average for Income table

The syntax for the ROWS clause is:

```
ROWS BETWEEN <start of frame> AND <end of frame>
```

Table 9-1 shows some expressions for specifying <start of frame> and/or <end of frame>. Remember that we always have to have an ORDER BY clause if we are using the ROWS clause.

Table 9-1. Specifying Rows of a Window

Expression	Meaning
UNBOUNDED PRECEDING	Start at the beginning of the current partition
<n> PRECEDING	Start n rows before the current row
CURRENT ROW	Can be used for either the start or end of frame
<m> FOLLOWING	End m rows after the current row
UNBOUNDED FOLLOWING	End at the end of the current partition

Here is the previous query with the (default) window of required rows spelled out:

```
SELECT Month, Area, Income,
    AVG(Income) OVER(
        PARTITION BY AREA
        ORDER BY Month
        ROWS BETWEEN UNBOUNDED PRECEDING AND CURRENT ROW
    ) AS AreaRunningAverage
FROM Income;
```

The ROWS clause in the preceding query is the default if no ROWS clause is specified after an ORDER BY clause.

Now we can change which rows are to be included in the average. Say we would like to see rolling three-month averages. This means that for each month we take an average that includes the current month, the one preceding, and the one following. The following query shows how we can add another ROWS clause to the preceding query to see both the running average and the rolling three-month average:

```
SELECT Month, Area, Income,
    AVG(Income) OVER(
        PARTITION BY AREA
        ORDER BY Month
        ROWS BETWEEN UNBOUNDED PRECEDING AND CURRENT ROW
    ) AS AreaRunningAverage,
    AVG(Income) OVER(
        PARTITION BY AREA
        ORDER BY Month
        ROWS BETWEEN 1 PRECEDING AND 1 FOLLOWING
    ) AS Area3MonthAverage
FROM Income;
```

Figure 9-13 shows the output of this query. The boxes show which values are contributing to the averages on the rows for month 4 (solid boxes) and month 9 (dashed boxes).

Month	Area	Income	AreaRunningAverage	Area3MonthAverage
1	Halswell	2400	2400	3100
2	Halswell	3800	3100	2533
3	Halswell	1400	2533	3233
4	Halswell	4500	3025	4033
5	Halswell	6200	3660	5167
6	Halswell	4800	3850	5100
7	Halswell	4300	3914	4667
8	Halswell	4900	4038	4800
9	Halswell	5200	4167	4933
10	Halswell	4700	4220	4800
11	Halswell	4500	4245	4733
12	Halswell	5000	4308	4750

Figure 9-13. Running averages and rolling three-month averages

The RunningAverage in the row for month 4 includes all the values from the beginning to month 4, and similarly the RunningAverage in the row for month 9 includes all the incomes up to and including month 9. The Rolling3MonthAverage in row 4 includes months 3 to 5 (one month preceding and one month following the current row). In row 9 the Rolling3MonthAverage averages months 8 to 10 (i.e., one month each side of month 9).

The different averages provide different information about the how the business is doing. The running average provides the average income to date for the year. The rolling three-month average gives a better idea of how the income is tracking at the moment. The later values in the rolling average column are higher than their running average counterparts because they are not including the lower values in the first few months.

Summary

Window functions provide an elegant way to carry out partitioning, running, and rolling aggregates and allowing both the detail and the aggregate to be available in the same query.

Here is a brief summary of the functionality covered in this chapter. I have used the word *table* in the descriptions but the functionality equally applies to the result of a query.

OVER()

Use the OVER() function with no clauses in the parentheses to calculate the aggregate for the whole table. Unlike simple aggregates it is possible to include other attributes in the SELECT clause, thereby retaining access to the detail as well as the aggregated value.

OVER(PARTITION BY <...>)

If PARTITION BY is included in the OVER() function then the rows are separated into groups that have the same value for the partitioning expression. The aggregates are carried out for each partition. This is similar to GROUP BY for a simple aggregate but has the advantage that several different partitions can be included in a single query.

OVER(ORDER BY <...>)

When ORDER BY is included in the OVER() function then the table is (virtually) ordered by the order by expression. The aggregate is then evaluated for the rows from the beginning of the table to the current row (and any following rows with the same value for the ordering expression). This is used for running aggregates.

OVER(PARTITION BY <...> ORDER BY <...>)

The table is first partitioned into different groups with the same value for the partitioning expression, and the rows are then ordered by the ordering expression within those groups. The aggregate is then evaluated for the rows from the beginning of the table to the current row (and any following rows with the same value for the ordering expression).

OVER(ROWS BETWEEN <...> AND <...>)

A ROWS BETWEEN clause can be added to an OVER() function with an ORDER BY clause. This restricts the aggregate to a set or rows relative to the current row, typically a number of rows preceding and or following the current row. It is useful for calculating rolling aggregates.

CHAPTER 10

■ ■ ■

Efficiency Considerations

You may not need to read this chapter! Database management systems (DBMS) are very efficient, and if you have a modest amount of data, most of your queries will probably be carried out in the blink of an eye. Complicating your life in an attempt to make your queries a little faster does not make a great deal of sense. However, if you have (or might have) vast amounts of data and speed is absolutely critical, you will need more skill and experience than you are likely to get from reading one chapter in a beginners' book. Having said that, you are likely to have people tell you that it matters how you express your queries or that you should be indexing your tables, so it is handy to have some idea about what is going on behind the scenes and understand some of the terminology.

Throughout this book, I have emphasized that there are often alternative ways to phrase a query in SQL. The implementation of SQL you are using may not support some constructions, so your choices may be limited. Even then, you usually have alternatives for most queries. Does it matter which one you use? One consideration is the transportability of your queries. If you are unsure where your query may be used, you might choose to avoid keywords and operations that are not widely supported (yet). However, typically you will be writing queries for a database with a specific implementation of SQL. In that case, your main questions are "How will the different constructions of a query affect the performance?" and "Is there anything I can do to improve performance?"

What Happens to a Query

Up to this point we have been concentrating on taking a question we need answered and constructing a query that will return appropriate and accurate information from the database. Conceptually the query writer thinks of the database as being a collection of tables. An SQL statement is an expression describing which data should be retrieved from those tables and what constraints that data must obey (the outcome approach).[1]

We have also seen that a query can be specified by describing set operations, such as joins and intersections, which would result in the appropriate data being returned (the process approach). Using set operations makes forming a query very elegant, but the operations are purely conceptual. While we might specify the query in terms of, say, a join followed by an intersection, this will be interpreted by the DBMS as a *description* of the data to be returned not as a *method* for retrieving the data.

The ideas of tables and data models is a useful way for us to understand how the pieces of data are logically related. We leave it to the DBMS to take care of how the data is physically stored and retrieved. Figure 10-1 is a simplified schematic of the different levels of abstraction that can help us understand a database.

---

[1]SQL is based on relational calculus, which provides a description of the data to be retrieved. See Appendix 2 for more information.

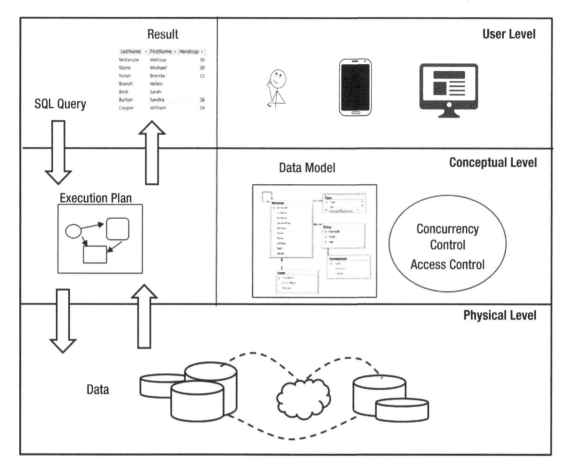

Figure 10-1. *Levels of a database management system*

At the top of the diagram in Figure 10-1 we have the user level. This is where we have the different applications and devices that form the interface between the humans and the data. It is here that a user (or application) will construct an SQL statement (top left of the diagram).

The middle layer is a conceptual view of the database. We can think of the data as being in tables with various key and validation constraints. It is also where we can form models of how the database should deal with concurrent users and the rules for allowing access to different data. The SQL statements we construct are in terms of all these concepts.

The actual data is stored in the physical level, shown at the bottom of Figure 10-1. What we think of as a table may be segments of data stored on possibly different servers maybe in different countries. At this level there will be indexes that allow rapid access to different records; we will talk about those in later sections of this chapter.

How the relevant pieces of information are located and assembled to produce the result of the query is a job for the *query optimizer*. The SQL statement constructed by the user is passed to the optimizer, which has access to information about the number of rows in a (conceptual) table, the amount of data in each row, the attributes on which indexes have been created, and so on. It uses all this information to create an *execution plan*. The execution plan is an efficient sequence of steps to find, compare, and assemble the data into the result specified by the query. The data is retrieved from physical storage and assembled in accordance with the execution plan, and the result is returned to the user – usually in the form of a table.

Most commercial database software provides tools for displaying proposed execution plans and the estimated times for each step to be carried out. This provides insight into how a query is being executed and where the time is being spent. A good database administrator will be able to use this type of information to tune the database; for example, by adding new indexes.

In the following sections, we will take a brief look at how records are stored, how indexes can improve efficiency, and some of the things that go on behind the scenes when a query is carried out.

Finding a Record

Most of the queries on a database will, at some point, involve finding records that match a particular condition. For example we may want to find records in a single table (e.g., WHERE LastName = 'Smith'), find records from two tables that match a join condition (e.g., WHERE m.MemberID = e.MemberID), or look for the existence or otherwise of values (e.g., WHERE NOT EXISTS...).

Searching for and finding data all seems pretty easy these days when everything is stored electronically. If we want to find a topic in an online book we just open a search box and type in some keywords. With a physical book it is a very different story. We either have to scan through every page or hope there is a useful index or table of contents. Those who remember physical telephone books will recall that it was easy to find Jim Smith but impossible to find who lives at 16 Murray Place.

Behind the scenes in a database the issues are the same as for physical books. The data can only be stored in one order, but we might want to search it in a variety of ways. It is useful to know what is actually going on so that we have an understanding of what affects the performance of locating a specific record.

One way to find records matching a condition is to simply look sequentially through every row in the table. This is the slowest and costliest way to find what you are looking for. Having said that, it may not be a problem. It would not take long to scan the golf club Type table to find the membership fee for seniors. On the other hand, it would be a bigger job to scan every row in (a realistic) Entry table to find members who had entered tournament 38 over the last forty years.

Storing Records in Order

If we consider relational theory, then the rows in a table[2] have no order. This allows us to consider a table as a set and apply all the set operations. This is useful from a conceptual point of view, but in practice how the records are stored is going to make a difference in how quickly we can find what we are looking for.

If the records of a table are stored in a random order (perhaps the order they were created) then this is referred to as a *heap table*. The only way to find a record in a heap table is to scan the entire table. Generally the records will be stored ordered by some attribute(s) – often the primary key. There are all sorts of algorithms for finding a particular record quickly in an ordered table, but most will be built on the idea of a *binary search*.

When we try to find a name in a telephone book or a word in a dictionary we employ a type of binary search. In the simplest scenario we inspect a page in the middle of the book and decide if the target word is before or after the words on that page. We then start again and inspect a page halfway through the portion that is of interest. Very quickly we zero in on the page required. If the records in a table are ordered by a particular field, then searches on that field will be more efficient than searches on fields with no index.

When we talk about records in a table being stored in order, we don't mean they are physically one after the other on a disc. If this were the case then if we wanted to insert a new record near the beginning we would have to move all the others along. The records can be thought of as being in a tree-like structure. One common type of tree is a *B-Tree*. Figure 10-2 shows a very simple representation of a B-Tree structure for storing letters of the alphabet.

---

[2]More formally the tuples in a relation have no order.

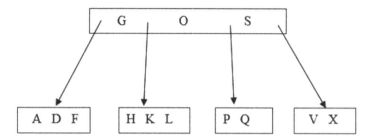

Figure 10-2. *Representation of data in a B-Tree*

If you are searching for a letter in the tree in Figure 10-2 you would start at the top node, or box. In the top node you either find what you are looking for or you follow the appropriate path. For example, if you are looking for H you would follow the path between G and O. The structure allows records to be added and deleted with the minimum of disruption. We can easily add R to the box containing P and Q without altering any of the other letters. Maintaining a B-Tree is not trivial. As data is added and removed the tree will need to be rearranged to keep it balanced and to add and remove nodes and levels. Fortunately, all this goes on under the hood.

Having said all that, when we are thinking about records being in order it is usually easiest to imagine them all in a single line. That is what I will do for the rest of this chapter.

Clustered Index

If the records are physically stored in some order then this is referred to as a *clustered index*. If we thought it a good idea to store our member records in order of names, we could specifically create a clustered index so that the records are stored in order of the value of LastName. We can create an index with an SQL statement. We need to provide a name for the index (e.g., Clustered_Name) and specify the field(s) on which to order the index, as in the query here:

```
CREATE CLUSTERED INDEX Clustered_Name ON Member (LastName);
```

By default, the order for a clustered index is usually the value of the primary key. While it is possible to specify a different order for the clustered index, you need to have a good reason to do so.

With a clustered index in place, there are now two ways to locate a record. Consider running the following query on the Member table with the clustered index on LastName:

```
SELECT *
FROM MEMBER
WHERE LastName = 'Smith';
```

Because the table is in order of LastName we can quickly navigate to the correct record by doing a binary search. This is known as a *table seek*.

Now consider the following query:

```
SELECT *
FROM MEMBER
WHERE Phone = '03-567-123';
```

We have no option now but to check every record in the table. We cannot even stop when we get to a matching record, as there may be several records with the same phone number. This is known as a *table scan*.

Non-Clustered Indexes

The records in a table can only be stored in one physical order, so there can only ever be one clustered index, which is usually on the primary key. If the clustered index for the Member table is on the primary key field MemberID, we can seek a particular value of MemberID and find the complete row with all the details for that member. If we want to find a row with a particular last name we would have to scan the whole table. Fortunately, we are able to set up additional non-clustered indexes on the table. I'll refer to these non-clustered indexes as simply indexes from now on. Here is the SQL to create an index on LastName for the Member table:

```
CREATE INDEX idx_Name ON Member (LastName);
```

What this does is create a list of all the values of LastName in order. Each entry will include a reference to the clustered index so that the full row with the rest of the information can be found. Figure 10-3 illustrates how the first few entries in the LastName index refer to the clustered index.

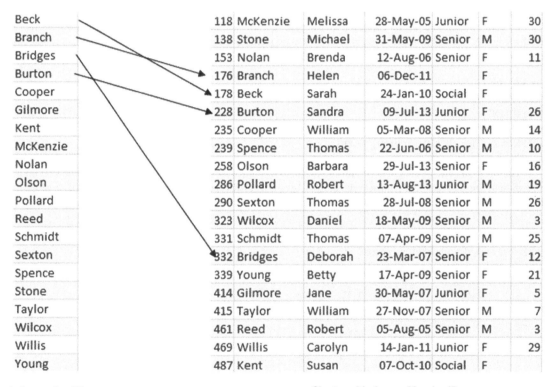

Index on LastName		Clustered Index on MemberID					
Beck	118	McKenzie	Melissa	28-May-05	Junior	F	30
Branch	138	Stone	Michael	31-May-09	Senior	M	30
Bridges	153	Nolan	Brenda	12-Aug-06	Senior	F	11
Burton	176	Branch	Helen	06-Dec-11		F	
Cooper	178	Beck	Sarah	24-Jan-10	Social	F	
Gilmore	228	Burton	Sandra	09-Jul-13	Junior	F	26
Kent	235	Cooper	William	05-Mar-08	Senior	M	14
McKenzie	239	Spence	Thomas	22-Jun-06	Senior	M	10
Nolan	258	Olson	Barbara	29-Jul-13	Senior	F	16
Olson	286	Pollard	Robert	13-Aug-13	Junior	M	19
Pollard	290	Sexton	Thomas	28-Jul-08	Senior	M	26
Reed	323	Wilcox	Daniel	18-May-09	Senior	M	3
Schmidt	331	Schmidt	Thomas	07-Apr-09	Senior	M	25
Sexton	332	Bridges	Deborah	23-Mar-07	Senior	F	12
Spence	339	Young	Betty	17-Apr-09	Senior	F	21
Stone	414	Gilmore	Jane	30-May-07	Junior	F	5
Taylor	415	Taylor	William	27-Nov-07	Senior	M	7
Wilcox	461	Reed	Robert	05-Aug-05	Senior	M	3
Willis	469	Willis	Carolyn	14-Jan-11	Junior	F	29
Young	487	Kent	Susan	07-Oct-10	Social	F	

Figure 10-3. *Index on LastName has references to clustered index for full information*

Because the index is ordered by last name it is possible to do an *index seek* to find the entry we require and then use the reference to *look up* the associated row in the clustered index to retrieve the rest of the information.

In practice, whether the query optimizer uses a particular index depends on many factors: the number of rows in the table, the size of each row in the index and the table, whether the records have been accessed recently and have been cached, and so on.

Clustered Index on a Compound Key

Let's consider the Entry table. Recall that the Entry table has three fields: MemberID, TourID and Year. Two questions we might ask about the data in this table are:

- Which tournament has a particular member entered (say, member 235)?

- Who has entered a particular tournament (say, tournament 40)?

It would seem sensible to have two indexes: one on TourID and one on MemberID. However, each of these would refer to the clustered index. What order will that be for the Entry table?

By default, the table will be clustered on the primary key, which for the Entry table is a combination of all three fields. The order of the records will depend on how we specified the primary key. Let's say the Entry table was created with the following SQL statement:

```
CREATE Table Entry (
MemberID INT,
TourID INT,
Year INT,
PRIMARY KEY (MemberID, TourID, Year);
```

The order of rows in the clustered index will be as in Figure 10-4. First, they are ordered by the first field specified in the PRIMARY KEY clause (MemberID). Those rows with the same value of MemberID will be ordered by the second field (TourID) and so on.

MemberID ⇥	TourID ⇥	Year ⇥
118	24	2014
228	24	2015
228	25	2015
228	36	2015
235	38	2013
235	38	2015
235	40	2014
235	40	2015
239	25	2015
239	40	2013

Figure 10-4. *Order of data in the default clustered index for the Entry table*

The system can easily find the tournaments which member 235 has entered because the entries are in order of `MemberID` and a table seek can be carried out. We do not necessarily need an additional index on `MemberID`. On the other hand, finding who has entered tournament 40 would require a scan to investigate every row. In this situation an index on `TourID` would be an improvement.

The order in which we specify the fields of the primary key can therefore affect how queries are carried out and can influence what other indexes might be useful. Had the order of the primary key fields been specified with `PRIMARY KEY (TourID, MemberID, Year)` then the clustered index would be in order of `TourID`. In that case, an index on `MemberID` should be considered if we regularly need to find rows for a particular member.

I was careful to say for the situation in Figure 10-4 that we might not *necessarily* need an index on `MemberID`. The optimizer will take into account many things. One that is important is the size or number of bytes of a typical entry in the index. Each entry in an index made up of two text fields such as `LastName` and `FirstName` will be larger than an index on a single text field, which will in turn be larger than an index on a numeric field. Each time an entry in an index is visited it has to be retrieved, so there is an IO (input/output) cost that will depend on the size of the entry.

If the clustered index in Figure 10-3 had significantly more data in each row (e.g., some descriptive text fields) then the cost of retrieving a row would be higher than for retrieving just the three numeric fields. In that case having an index on `MemberID` would be worth considering. It would not alter how many index entries needed to be investigated, but it would have a smaller IO cost as each entry is smaller. The downside is that once the correct `MemberID` is located by an index seek the system will need to look up the clustered index to find the rest of the information. Depending on all the information it can access, the optimizer will determine whether it is more efficient to use the index on `MemberID` and look up the rest of the information, or just to scan all the records in the clustered index.

Updating Indexes

Indexes are clearly wonderfully useful. Why do we not just index everything we are ever likely to search on? This is certainly possible. The downside is that the indexes have to be maintained. Every time we add or delete a record in a table every index on that table will need to be updated also. We therefore have a tradeoff. Lots of indexes will mean fast retrieval but slower updating. Fewer indexes will mean faster updating but possibly slower retrieval.

Managing these tradeoffs is work for an experienced database administrator with excellent knowledge of the domain. There are many tools available that will monitor the database and provide statistics on the use of indexes and other information about the data. If the data is relatively stable with few updates then having several indexes will make retrieval faster. If the data is constantly being updated then indexes may be counterproductive.

In situations where there are a lot of updates it may be practical to do bulk updates of data. With a bulk update you can remove the indexes. The following query shows how to remove the index we created on the `Member` table earlier in the chapter:

```
Drop idx_Name on Member;
```

All the additions, deletions, and modifications to the table can then be carried out without the overhead of updating the indexes. At the end of the updates on the table, the indexes can be recreated. This may or may not be more efficient than updating each index for every change.

Covering Indexes

Adding more fields to some indexes can also be effective. Consider the following query:

```
SELECT FirstName
FROM Member
WHERE LastName = 'Smith'
```

If the Member table has an index on LastName, then the preceding query would require an index seek to find Smith and then a lookup of the clustered index to find the first name. If the index was on the compound index (LastName, FirstName) then all the information required for the query is contained in the index and no lookup is required. This is known as a *covering index*. Again, there is the tradeoff of having a larger IO cost for the bigger rows in the index versus the cost of the lookup.

Selectivity of Indexes

Indexes are most useful when the number of rows returned by an index search is small compared to the number of rows in the table.

For example, let's consider finding information about a member with a particular last name. An index on LastName in the Member table is likely to return only a small percentage of its elements if we search for a specific name. The system can then look up the clustered index for each of those returned elements to retrieve the rest of the information about the members.

By contrast, what happens if we want to find information about women. An index on Gender will return around half its entries if we search for 'F' in the golf club's Member table. The DBMS would then have to look up the corresponding records in the clustered index. In this situation it is probably more efficient to just scan the clustered index, which contains all the information we require, and not bother with the index at all.

Sometimes the selectivity of an index is not obvious. For example, an index on a field City will not be useful if most of the records in the table have the same value for city and most of the queries are for that city.

Database software often provides tools that can help us. The tools might provide statistics on the current spread of data in fields in a table – for example, what percentage of the table has the same values in a field, such as City. This will help determine if an index might be useful. Often statistics can be collected about how often an index is used. If the optimizer makes little use of an index then it might as well be removed rather than be constantly updated.

Join Techniques

If we consider the Entry table in Figure 10-4, most queries will require a join on the Member table to find the names of the entrants and/or a join on the Tournament table to find the names and other information about the tournaments. Each of these joins compares a foreign key in the Entry table with the primary key of the Member or Tournament table. Refer to Chapter 1 to review what we mean by a foreign key. This joining of a foreign key with a primary key is such a common scenario that it is worth understanding how joins can be carried out. We will use the Member and Entry tables as an example, but the ideas have wide application.

There are a number of different approaches that can be taken when carrying out a join. Which approach will be the most efficient will depend on many things, including the relative sizes of the tables, the indexes that have been created, whether the query also includes projecting specific columns or selecting rows, whether an output order has been specified, and so on. You don't have to worry about the choice of approach, as that will be decided by the optimizer. However, creating particular indexes can influence the approach taken.

Nested Loops

One approach to joining tables is called *nested loops*. This means that the system scans one table, and for each row in that table looks through all the rows in the other table to find matches for the join condition. The nested-loop approach is illustrated in Figure 10-5 for the join condition `Entry.MemberID = Member.MemberID`.

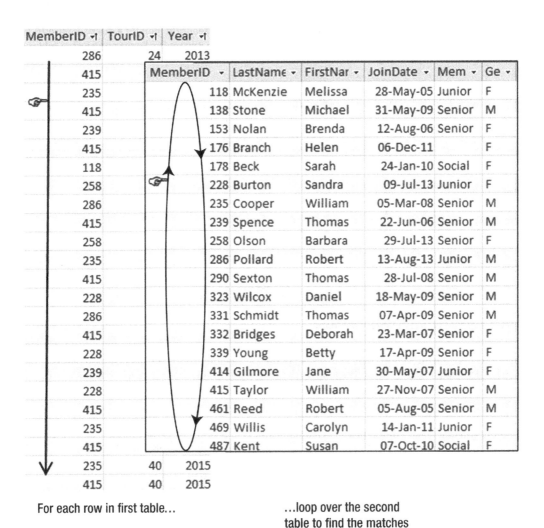

For each row in first table...   ...loop over the second
                                                   table to find the matches

Figure 10-5. *Nested-loops approach to finding rows with matching MemberIDs*

In Figure 10-5 the outside loop is on the `Entry` table. For each row in the `Entry` table, the system will need to find the matching rows in the `Member` (inner) table. The tables shown in Figure 10-5 are not ordered, which means that every row of the `Member` table will need to be visited for each row in the `Entry` table to find the matching records (a table scan).

If there is an index on the matching field in the inner loop (MemberID in the Member table, in this case) then finding the matching field will be more efficient. The index can be used to quickly find the matching records without having to visit every row. In practice, the Member table will probably have a clustered index on its primary key MemberID. If the tables are nested with the Entry table on the inside, then the internal loop will be more effective if there is an index on the MemberID of the Entry table. The optimizer will take this information into account to decide if the nested-loops option is efficient for carrying out the join and which tables should be the inner and outer tables.

Most commercial database systems will provide tools to view the execution plan for a query. Figure 10-6 shows a screenshot from SQL Server showing the execution plan for the join on the Member and Entry tables in the following query:

```
SELECT *
FROM Member m INNER JOIN Entry e on m.MemberID = e.MemberID;
```

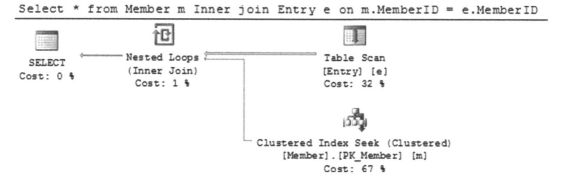

Figure 10-6. *Execution plan showing nested loops*

In Figure 10-6 we see on the top right a table scan of the Entry table. This is the outside loop of the nested loop (as depicted in Figure 10-5). The icon on the bottom right shows that for each row in the Entry table, a seek on the clustered index of the Member table will be carried out to find the row with a matching MemberID.

Does it matter in which order we specify the tables in a join query? If we put the Entry table first in the SQL expression, will that make a difference? Once upon a time it may have. These days almost certainly not. Expressing the query with the table in a different order results in the same execution plan in SQL Server as the plan in Figure 10-6. However, if we change which fields are being selected, or add other indexes, or choose to sort the output, then the execution plan will very probably change.

Merge Join

Another approach to doing a join is to first sort both tables by the join field. It is then very easy to find matching rows. This is called a *merge join* and is shown in Figure 10-7.

MemberID	TourID	Year
118	24	2014
228	24	2015
228	25	2015
228	36	2015
235	38	2013
235	40	2014
235	38	2015
235	40	2015
239	40	2013
239	25	2015
258	24	2014
258	38	2014

MemberID	LastName	FirstNar
118	McKenzie	Melissa
138	Stone	Michael
153	Nolan	Brenda
176	Branch	Helen
178	Beck	Sarah
228	Burton	Sandra
235	Cooper	William
239	Spence	Thomas

Figure 10-7. *Merge join requires each table to be sorted by the field being compared*

Sorting tables is an expensive operation. However, if the tables have indexes on the join fields then the rows can be accessed in order by an index scan, making the merge join an option.

Both the merge join and the nested-loops join will be more effective if one or both of the fields in the join condition have indexes.

Different SQL Expressions for Joins

In the previous section I briefly touched on whether the order of the tables in a join would affect the execution. The answer was no for the query in Figure 10-6. However, we have other ways of expressing joins. The two queries that follow have exactly the same execution plans in SQL Server:

```
SELECT LastName FROM Member m, Entry e WHERE m.MemberID = e.MemberID;

SELECT LastName FROM Entry e INNER JOIN Member m ON m.MemberID = e.MemberID;
```

The following two SQL statements specify the join in terms of nested queries. They have different execution plans from the preceding queries but they are the same as each other:

```
SELECT LastName FROM Member m WHERE m.memberID IN
    (SELECT MemberID FROM Entry);

SELECT LastName from Member m WHERE EXISTS
  (SELECT * FROM Entry e WHERE m.MemberID = e.MemberID);
```

So, should we use or avoid nested queries? The answer, as always, is "it depends."

171

Before we compare the preceding two pairs of SQL statements we need to be aware that the output from them is different. The first pair will produce duplicate names (repeating a member's name for every tournament they have entered). The second pair of queries will produce unique names. To be fair in the comparison we will compare the following two queries, which have identical output:

```
SELECT DISTINCT LastName
FROM Entry e INNER JOIN Member m ON m.MemberID = e.MemberID;

SELECT LastName FROM Member m WHERE m.MemberID IN
    (SELECT MemberID FROM Entry);
```

You can see the two plans in Figure 10-8.

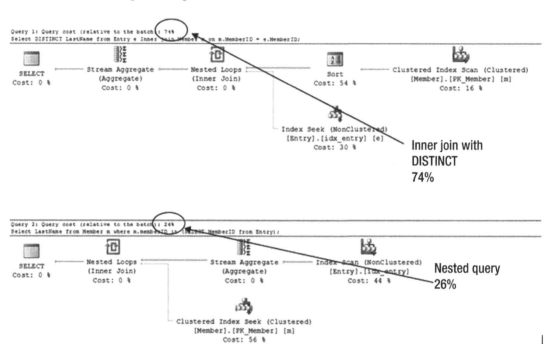

Figure 10-8. *The same output but very different execution plans and costs*

Figure 10-8 shows the plan for the query using the INNER JOIN keyword at the top and the plan for the nested query underneath. The percentages are saying that if both these queries were executed in one batch then the top one would account for 74 percent of the time and the bottom one 26 percent. That is, the INNER JOIN query takes three times as long as the nested query.

The addition of the DISTINCT keyword in the top query accounts for much of the time. The optimizer has chosen to sort the records in order to prepare to remove the duplicate names. This sorting operation accounts for over half the total cost of the first query. Seeing this plan, you might consider adding an index on LastName so that the records for the Member table could be accessed in LastName order, thus eliminating the need for the time-consuming sort.

Unless you have real insider knowledge, it is just about impossible to second guess what the optimizer will come up with. In the long run it probably doesn't matter unless the tables have huge numbers of rows or a query is particularly time critical. The important thing to remember is that if you suspect that a critical query is causing a bottleneck, there are tools that can help you understand what is going on. You can then experiment with indexes or the ways the query is expressed to see if that can speed things up.

Summary

Indexes can make a considerable difference in the performance of many queries. However, there is the downside that they have to be maintained. With any tuning of a database it is important to know what the important processes are. There is no point going to any lengths to improve a query that is rarely carried out, and it is counterproductive to improve retrieval performance if most of the time-critical work in your database is the updating of records.

The tools provided by many database systems can provide valuable information. Execution plans can give insight into where the time is being spent in a query. Statistics can be collected about the use of indexes or the distribution of data in a field. All this information is useful when deciding whether the addition of a new index might be worth investigating.

Here are some general rules of thumb for creating indexes.

Primary Key

You need a very good reason not to have an index on the primary key field(s) of a table. Generally a clustered index will be placed on the primary key by default.

Foreign Keys

Joins where the join condition is between a foreign key and a primary key are very common. For this reason an index on foreign keys is usually worth considering.

WHERE Conditions

If you have queries that frequently use particular fields in a WHERE condition, then it is useful to index on those fields. This enables an index seek rather than having to do a table scan to find the relevant rows. This is most useful when the WHERE condition is selective, meaning that it will retrieve only a small subset of the rows.

ORDER BY, GROUP BY, and DISTINCT

Sorting can be a very expensive operation if there are no indexes on the fields involved in the sorting condition. Clearly ORDER BY requires rows to be sorted. Queries that contains DISTINCT or GROUP BY often sort the records to remove duplicates or to aggregate the data. With appropriate indexes, an index scan can be used to retrieve the rows in order, thus eliminating the need for an expensive sorting operation.

Use the Tools

Query optimizers are very sophisticated. They maintain statistics about your tables (number of rows, size of columns, distribution of data, etc.) and use these to help determine an efficient execution plan for a query. If you have a critical query that you want to be as efficient as possible, check the execution plans to see where the time is being spent. You can then experiment with the effects of restating the query or adding additional indexes.

CHAPTER 11

■ ■ ■

How to Tackle a Query

In the previous chapters, we saw different ways to express a query. We looked at the *process approach*, which describes how tables and data could be manipulated to produce the required result. These queries are expressed using keywords describing operations such as INNER JOIN and INTERSECTION. We also looked at how to express queries in terms of the *outcome approach*, which describes the criteria that the resulting data must satisfy rather than the process for retrieving the result.

However, sometimes when I am presented with a complicated natural language description of a query, it is not uncommon to find that my mind goes blank. I have a lot of ammunition at hand, but for a moment or two, have no idea which weapons to choose.

Usually, it is just a matter of being confident and relaxed. Large, complicated queries can always be broken down into a series of smaller, simpler queries that can be combined later. This chapter describes how to do just that.

Understanding the Data

It may sound like stating the obvious, but you can't retrieve information from a database without understanding where all the different elements of data are stored and how the relevant tables are interrelated. Most of the time you will be querying a database designed by someone else, and probably maintained and altered over time by various people. As well as understanding the tables and relationships that have been implemented, it is also necessary to have a feel for the underlying real-world scenario. You also must be alert to the unfortunate reality that the database may have been badly designed. This might mean that you are not able to retrieve the required information accurately. We will consider this problem of working against bad design a bit more in Chapter 12.

Determine the Relationships Between Tables

The best way to get an overview of the implementation of a database is to look at a schematic of the relationships between the tables. Most database management software provides a way of viewing the fields in the tables and the foreign key relationships between the tables. Figures 11-1 and 11-2 show the foreign key relationship diagrams for our club database as depicted by SQL Server and Microsoft Access.

© Clare Churcher 2016
C. Churcher, *Beginning SQL Queries*, DOI 10.1007/978-1-4842-1955-3_11

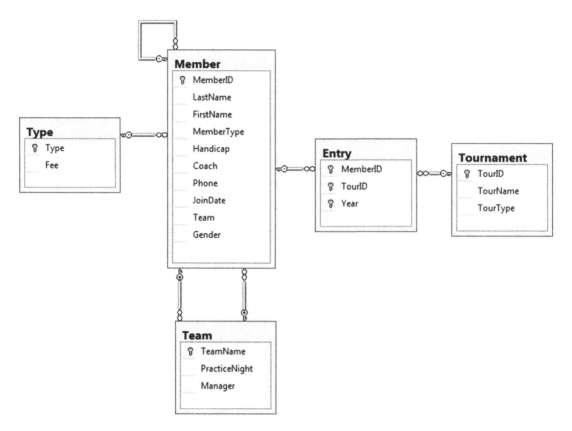

Figure 11-1. *A database diagram from SQL Server*

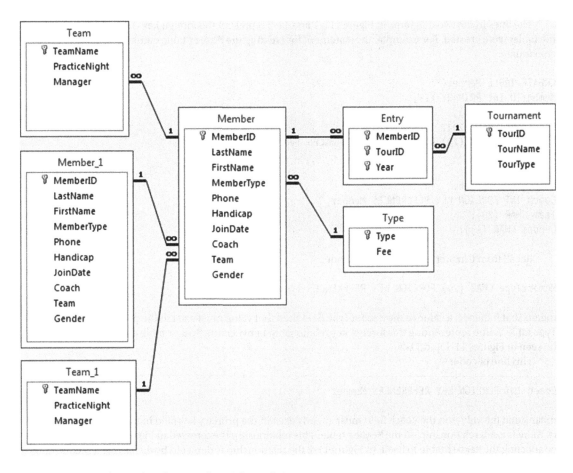

Figure 11-2. *A relationship diagram from Microsoft Access*

On the surface, the diagrams in Figures 11-1 and 11-2 look a bit different, but they represent exactly the same database. The Access schematic in Figure 11-2 displays an additional copy of the Member and Team tables. The two copies of the Member table arise from the self relationship between members (that is, a member can coach other members). The additional copy of the Team table is because of the two relationships between Member and Team: a member can be the *manager* of a team, and a member can *belong* to a team. These relationships are depicted in the SQL Server diagram in Figure 11-1 by showing two lines between the tables so that the tables are not shown twice. The different diagrammatic representations are just a quirk of the different management systems. Both schematics represent the same set of tables and relationships.

The lines in the two diagrams in Figures 11-1 and 11-2 represent the foreign keys that were set up when the tables were created. For example, the statement for creating the Member table contains two foreign key constraints:

```
CREATE TABLE Member(
MemberID Int PRIMARY KEY,
LastName CHAR(20),
FirstName CHAR (20),
MemberType CHAR (20) FOREIGN KEY REFERENCES Type,
Phone CHAR (20),
Handicap INT,
JoinDate DATETIME,
Coach INT FOREIGN KEY REFERENCES Member,
Team CHAR (20),
Gender CHAR (1));
```

Recall from Chapter 1 that this line of code:

```
MemberType CHAR (20) FOREIGN KEY REFERENCES Type
```

means that if there is a value in the MemberType field then that value must exist in the primary key field in the Type table. A line representing this foreign key relationship between the Member table and the Type table can be seen in Figures 11-1 and 11-2.

This line of code:

```
Coach INT FOREIGN KEY REFERENCES Member
```

means that the values in the Coach field must already exist in the primary key field in the Member table; that is, there is a self relationship on the Member table. This relationship is expressed in Figure 11-1 with the loop connecting the Member table to itself. In Figure 11-2 the relationship is depicted by displaying a second copy of the Member table.

Real World Versus Implementation

The database diagrams in the previous section represent how the database has been *implemented* and in particular which foreign keys have been set up. When the database is first set up, the design will be based on a *conceptual* data model that describes how the tables for a particular problem are interrelated. A number of methods exist for representing a data model, such as entity-relationship (ER) diagrams and the UML class diagrams we have been using in this book. Figure 11-3 shows the class diagram for the golf club. Refer back to Chapter 1 if you need a refresher on how to interpret the lines and numbers.

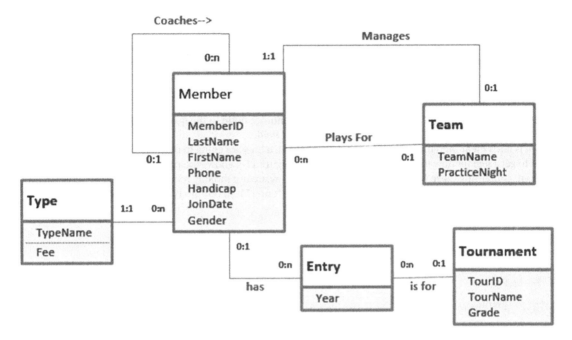

Figure 11-3. *Class diagram representing the conceptual model*

The class diagram in Figure 11-3 does not display foreign key fields in the classes. You can see this by comparing the Entry table in Figure 11-3 with those in the two earlier database diagrams. The foreign keys MemberID and TourID are missing as attributes in the class diagram. Foreign keys are simply a way of representing the relationship between classes if we choose to implement the data model in a relational database. If we decide to implement it in an object-oriented database we might not need foreign key fields at all.

A class diagram with well-labelled relationships gives us a much greater understanding of the real-world situation than do the implementation diagrams in Figures 11-1 and 11-2. Have a look at the relationships between Member and Team to see what I mean.

The database diagrams presented by relational database software show you the foreign keys that have actually been set up. These may not tell the whole story. The developer may not have implemented the relationship for coaching (for example) with a foreign key constraint on the Coach field. He or she may have overlooked the requirement or may have decided to enforce the constraint that a coach must be an existing member some other way (with a trigger or via the interface). However, even if there is no foreign key constraint on the Coach field in the Member table, we still need to understand that members coach other members if we want to design reliable queries about coaching.

In some cases, the implemented database may not have much in common with an accurate data model. For example, if the golf club database contained separate tables for members, coaches, and managers or one of the relationships between the Member and Team tables was not implemented, then the database diagram and the data model would look quite different. The likelihood of getting reliable information would be low. Chapter 12 looks at problems like this, although short of a major redesign there is sometimes not much you can do.

What Tables Are Involved?

Once we have an understanding of the tables in the database and how they are related (conceptually as well as by the existence of foreign keys), we can look at which tables you will need in order to extract the subset of data required. Consider a query like "Find all the men who have entered a Leeston tournament." This sentence contains a few key words. Nouns are often a clue to what tables or fields we are going to need. Verbs often help us find relationships. Let's look at the nouns. "Tournament" is a big clue, and we have a Tournament table, so that is a start. The word "men" is another noun in the query description. We don't have a Men table, but we do have a Member table with a Gender field.

It is fairly clear then that the Member and Tournament tables are going to play a part in our query. Now we need to get a feel for how these two tables are related. Figure 11-4 shows the part of the SQL Server database diagram containing these two tables. We see that that they are not directly related, but rather are connected via the Entry table. That makes sense, because the verb "enter" is in our query description.

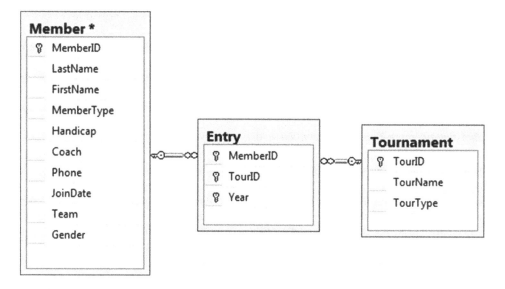

Figure 11-4. *Part of the database diagram showing the Member and Tournament tables*

So, it looks like at least three tables will be involved in our query: Member, Tournament, and Entry. We then use our understanding of the relational operators to determine how these tables could be combined. Do we need a join or a union, or some combination of these and other relational operators? We'll look at ways to help decide on the appropriate operations in later sections in this chapter.

Look at Some Data Values

Requests for information from a database are usually couched in rather informal and imprecise natural language. Even a simple request, such as "Find all the men who have entered a Leeston tournament," has a few things we need to clarify. Having a look at the actual data in the tables can sometimes help.

Our query does not actually "find" the men, but rather returns some information about them. Looking at the data values in the table will help us decide what information might be helpful. Presumably, the questioner would like to see the names of the men. Do we need the IDs as well? We will need IDs if we want to distinguish two members with the same name.

It may not always be clear what some of the words in the question refer to. What is a "Leeston" tournament? Is Leeston the name of a tournament, a type of tournament, or a location? Looking at a few rows of the Tournament table can help us. We see that the TourName field has the value "Leeston" here and there. Sometimes it might not be so easy to determine what imprecise words in the query description refer to. It may be necessary to talk to the developer or users to get a better understanding of what information they are trying to retrieve.

How do we determine which members are men? Fortunately, the Member table has a Gender column, and it looks like we want values of M. Is selecting rows with values of M going to be enough? Might there be some rows that have m or Male as the values? We'll look at how to deal with issues of inconsistent data in the next chapter. For now, let's assume that men are denoted by M.

For the simple query in this example, we now have a more precise description. It is something like "Retrieve the MemberID, LastName, and FirstName of the men (Gender = 'M') who have entered the tournament where TourName = 'Leeston'."

You might think of some other particulars that need clearing up. It is often a good idea to ask *why* this information is required. Do we just want to find which men have ever been to Leeston (because we want to ask one of them some questions about the golf course), or do we want to know how many times our male club members have entered Leeston tournaments (because we are interested in how popular the tournament is with the members of the club)? These questions can have different answers, as you will see in the "Retain the Appropriate Columns" section coming up soon.

Big Picture Method

My first attempt at a query is seldom elegant or complete. For a query such as "Find all the men who have entered a Leeston tournament," there are two ways I might tackle it, depending how my muses are working. One way is the big picture. I do this if I have a bit of an idea of how to combine the tables. I will cover another tactic in the section "No Idea Where to Start?", which I use when I have no idea where to start!

In the big picture method, I like to combine all the tables I'll need and retain all the columns, so I can see what is happening. I usually find it easiest to have an SQL window of some sort open so I can try small queries to see if the intermediate results look promising for answering the overall question.

Let's look at the big picture approach to the query "Find all the men who have entered the Leeston tournament." We decided we needed three tables: Member, Entry, and Tournament. These tables are all connected by foreign keys, and this often suggests that joins will be useful. If it isn't clear to you that a join is what is required for the query, then resort to the methods in the "No Idea Where to Start?" section later in this chapter.

Combine the Tables

Let's assume that we think joining tables looks like a promising approach for the query about men entering the Leeston tournament. You don't have to do everything at once. Start slowly with some small queries to see how things shape up.

To carry out a join, we need to find the fields on which to join. Review Chapter 3 if you need to refresh your understanding of join-compatible fields. The Entry table is critical to this query, as it connects the Member and Tournament tables. The Entry table has a foreign key field labeled TourID, which we can join with the primary key of the Tournament table. Do that much first.

```
SELECT * FROM
Tournament t INNER JOIN Entry e ON t.TourID = e.TourID;
```

Figure 11-5 shows a few rows of the resulting virtual table.

t.TourID	TourName	TourType	MemberID	e.TourID	Year
24	Leeston	Social	118	24	2014
24	Leeston	Social	228	24	2015
25	Kaiapoi	Social	228	25	2015
36	WestCoast	Open	228	36	2015
38	Canterbury	Open	235	38	2013
38	Canterbury	Open	235	38	2015
40	Otago	Open	235	40	2014
40	Otago	Open	235	40	2015
25	Kaiapoi	Social	239	25	2015
40	Otago	Open	239	40	2013
24	Leeston	Social	258	24	2014

Figure 11-5. Part of the result of joining the Tournament and Entry tables

The result shown in Figure 11-5 is certainly helpful. We can see the entries and the names of the corresponding tournaments. We can see from the first two rows that members 118 and 228 have entered a Leeston tournament. Now we need to find out whether 118, 228, and other members entering the tournament are men and find their names. We can get this additional information by joining the virtual table in Figure 11-5 to the Member table on the MemberID fields:

```
SELECT * FROM
(Tournament t INNER JOIN Entry e ON t.TourID=e.TourID)
INNER JOIN Member m ON m.MemberID = e.MemberID;
```

Figure 11-6 shows the result. I haven't included all the columns in Figure 11-6 because there are a lot of them. You will see shortly why I like to leave all the columns in as long as possible.

TourName	TourType	e.MemberI	e.TourID	m.Member	LastName	FirstName	Gender
Leeston	Social	118	24	118	McKenzie	Melissa	F
Leeston	Social	228	24	228	Burton	Sandra	F
Kaiapoi	Social	228	25	228	Burton	Sandra	F
WestCoast	Open	228	36	228	Burton	Sandra	F
Canterbury	Open	235	38	235	Cooper	William	M
Canterbury	Open	235	38	235	Cooper	William	M
Otago	Open	235	40	235	Cooper	William	M
Otago	Open	235	40	235	Cooper	William	M
Kaiapoi	Social	239	25	239	Spence	Thomas	M
Otago	Open	239	40	239	Spence	Thomas	M
Leeston	Social	258	24	258	Olson	Barbara	F
Canterbury	Open	258	38	258	Olson	Barbara	F
Leeston	Social	286	24	286	Pollard	Robert	M
Leeston	Social	286	24	286	Pollard	Robert	M
Leeston	Social	286	24	286	Pollard	Robert	M
Leeston	Social	415	24	415	Taylor	William	M
Kaiapoi	Social	415	25	415	Taylor	William	M

Figure 11-6. Part of the result of joining the Tournament, Entry, and Member tables (just some columns)

The virtual table shown in Figure 11-6 has all the information we need to find the required data. The first two rows show that members 118 and 228 are women. The row for member 286 (with the circles) looks more promising. How do we amend the query to find the appropriate subset of rows and columns?

Find the Subset of Rows

From Figure 11-6 we can see that the rows that we want to retain from the result of the join are where the Gender field has the value M and the TourName field has the value Leeston. We can select these rows by adding an appropriate WHERE clause to the previous query:

```
SELECT * FROM
(Entry e INNER JOIN Tournament t ON t.TourID=e.TourID)
INNER JOIN Member m ON m.MemberID = e.MemberID
WHERE m.Gender = 'M' AND t.TourName = 'Leeston';
```

Figure 11-7 shows just some of the columns from the result of the query above. It has four rows: three for Robert Pollard and one for William Taylor.

TourName	TourType	e.MemberID	e.TourID	Year	LastName	FirstName	Gender
Leeston	Social	286	24	2013	Pollard	Robert	M
Leeston	Social	286	24	2014	Pollard	Robert	M
Leeston	Social	286	24	2015	Pollard	Robert	M
Leeston	Social	415	24	2015	Taylor	William	M

Figure 11-7. *Men who have entered Leeston tournaments (just some columns)*

Why do we have three rows for Robert Pollard? The rows are identical except for the value of the Year field. Robert has entered the Leeston tournament in three different years. We can see this quite clearly from Figure 11-6 because we have left the Year column in the output. Had we retained only the name columns, we might initially be a bit puzzled at having Robert Pollard repeated three times. What we do about the repetition of Robert Pollard depends on understanding the initial question a bit more clearly, as you will see in the next section.

Retain the Appropriate Columns

We have the appropriate subset of rows from our large join. Now we need to retain just the columns we require by amending the SELECT clause, which is currently returning all the columns (SELECT *). This is not always as simple as it might sound. The three rows for Robert Pollard give us a bit of a clue that things may not be as straightforward. We have two possibilities.

If we only want to know who has entered the tournament in any year, then we want just the distinct names Robert Pollard and William Taylor and perhaps their ID numbers. Amending the SELECT clause as in the following query will provide that outcome:

```
SELECT DISTINCT m.MemberID, m.LastName, m.FirstName
FROM ...
```

If the objective of the question is to find out how often men enter Leeston tournaments, then we want to retain all the entries. In that case, it might be useful to retain the year as well to distinguish the rows as in the following:

```
SELECT m.MemberID, m.LastName, m.FirstName, e.Year
FROM ...
```

Consider an Intermediate View

The SQL for the joining the `Entry`, `Member,` and `Tournament` tables is likely to be the basis of many queries about entries in tournaments. For example, the following questions will all require a join of the `Member`, `Entry`, and `Tournament` tables:

- Do junior members enter open tournaments?

- Which tournaments did William Taylor enter in 2015?

- What is the average number of Social tournaments that members entered in 2013?

As we are likely to use this large join many times, it can be convenient to make a *view*. A view is an instruction for how to create a temporary table that we can use in other queries. The following is a first attempt at the SQL for creating a view that retains all the fields from the joins:

```
--First Attempt (unsuccessful)
CREATE VIEW AllTournamentInfo AS
SELECT * FROM
(Entry e INNER JOIN Tournament t ON t.TourID=e.TourID)
INNER JOIN Member m ON m.MemberID = e.MemberID;
```

As it stands, this query will not run in most versions of SQL. This is because the view would have fields with the same name; for example, there will be two fields called `MemberID`: one from the `Entry` table and one from the `Member` table.

When you create a view, all the field names must be distinct. The view will not use the aliases to differentiate the columns in the resulting table. The * in the `SELECT` clause needs to be altered to list all the field names. We need to either include just one of the fields with duplicated names (`MemberID` and `TourID`) or rename those that are duplicated (e.g., `SELECT m.MemberID AS MMember, e.MemberID AS EMember`). This is a bit tedious, but if you are creating a view that you are likely to use many times, it is worth the effort.

Once we have the view `AllTournamentInfo`, it can be used in the same way as any other table in our queries. To find the names of men who have entered a Leeston tournament, we can use the view as shown here:

```
SELECT DISTINCT LastName, FirstName
FROM AllTournamentInfo
WHERE Gender = 'M' AND TourName = 'Leeston';
```

Spotting Keywords in Questions

The big picture approach assumes that we have decided how to combine the tables that will contribute to the query. Sometimes, it will be obvious that, for example, certain tables need to be joined. Other times, it may not be at all clear initially. In this section, we will look at some keywords that often appear in questions and that can provide a clue about which relational operations are needed. If none of these help, remember that we still have the "No Idea Where to Start?" section coming up!

And, Both, Also

And and *also* are words that can be misleading when it comes to interpreting queries, and we will consider this further in the next chapter. In this section, we will look at queries that have the idea of two conditions needing to be met simultaneously. Queries that require two conditions to be met fall into two categories: those that can be carried out with a simple WHERE clause containing a Boolean AND operator, and those that require an intersection or self join.

To decide if a query really needs two conditions to be met, I usually look at a natural language statement and see if I can reword it with the word *both* connecting the conditions. Consider these examples:

- Find the junior boys. (*Both* a male and a junior? Yes.)

- Find those members who entered tournaments 24 *and* 38. (*Both* tournaments? Yes.)

- Find the women *and* children. (*Both* a female and a child? No.)

The last query is the one that can sometimes trip people up. Although it contains the word *and*, the common interpretation of "women and children" doesn't mean someone who is *both* a female and a child (that is a girl). Rather, the phrase means anyone who is *either* a female *or* a child (especially when populating lifeboats).

The diagram in Figure 11-8 is a useful way to visualize whether the natural language word *both* really means both or either. The circles represent the two sets: woman and children. Figure 11-8a shows the union (only one condition must be satisfied) and Figure 11-8b the intersection (both conditions must be satisfied).

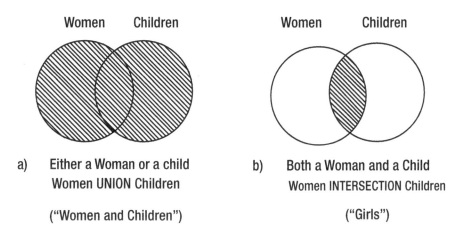

Figure 11-8. *Visualizing whether a union or an intersection is needed*

When two conditions must be met, we are looking at the intersection of two groups of data, as in Figure 11-8b. This doesn't necessarily mean we must use the INTERSECT keyword. I find the following question helpful in deciding what to do next:

> *Do I need to look at more than one row to decide if both conditions are satisfied?*

Consider the query to find junior boys. This is going to need the Member table. Can we look at a single row and determine if the member is *both* a junior and a boy? We can see in Figure 11-9 that both pieces of information are available in a single row.

MemberID ▾	LastName ▾	FirstName ▾	MemberType ▾	Gender ▾
178	Beck	Sarah	Social	F
228	Burton	Sandra	Junior	F
235	Cooper	William	Senior	M
239	Spence	Thomas	Senior	M
258	Olson	Barbara	Senior	F
286	Pollard	Robert	Junior	M
290	Sexton	Thomas	Senior	M
323	Wilcox	Daniel	Senior	M
331	Schmidt	Thomas	Senior	M
332	Bridges	Deborah	Senior	F

Figure 11-9. *Information about membership type and gender are available in a single row*

In this situation, we can use a simple SELECT operation with the Boolean AND to check for both conditions, as discussed in Chapter 2:

```
SELECT * FROM Member m
WHERE m.Gender = 'M' AND m.MemberType = 'Junior';
```

Now consider a different type of query. What about finding the members who have entered *both* tournaments 24 and 36? To do this, we need to look at the Entry table (probably joined with the Member table if we want the names). As we can see in Figure 11-10, we cannot check that a member, e.g., member 228, has entered both tournaments by looking at a single row.

MemberID ▾	TourID ▾	Year ▾
118	24	2014
228	24	2015
228	25	2015
228	36	2015
235	38	2013
235	38	2015
235	40	2014
235	40	2015

Figure 11-10. *We need to investigate more than one row to check both tournaments*

Where we need to satisfy both of two conditions *and* we need to look at more than one row in the table, we can either use a self join (discussed in Chapter 5) or an intersection (discussed in Chapter 7).

If we use the self join then the query is:

```
SELECT DISTINCT e1.MemberID
FROM Entry e1 INNER JOIN Entry e2 ON e1.MemberID = e2.MemberID
WHERE e1.TourID = 24 AND e2.TourID = 36;
```

A query producing the same output but using the INTERSECT keyword is:

```
SELECT MemberID FROM Entry WHERE TourID = 24
INTERSECT
SELECT MemberID FROM Entry WHERE TourID = 36;
```

Not, Never

Here are some examples of queries involving the words *not* or *never*:

- Find the members who are *not* seniors.

- Find members who are *not* in a team.

- Find members who have *never* entered a tournament.

Often when people see *not* in a description of a query, they immediately think of using a Boolean NOT or a <> operator in a WHERE clause. This is fine for some queries, but will fail for others. As in the previous section, I find the following test helpful to understand the category of the query.

Do I need to look at more than one row to decide if a condition is not true?

For the first two queries in the preceding bulleted list, we can look at a single row in the Member table and decide whether that member satisfies the condition. In the first query, the condition in the WHERE clause would be NOT MemberType = 'Senior' or MemberType <> 'Senior'. To find members who are not in a team, we want the Team field to be empty, so a clause like WHERE Team IS NULL would do the trick.

To find the members who have never entered a tournament, what tables do we need? We are certainly going to need the Entry table. We can decide if a member has entered a tournament by finding just one row with his or her value of MemberID. To see if he or she has *not* entered a tournament, we need to look at every row in the Entry table. We also must look at the Member table, because those members who have not entered a tournament will not appear in the Entry table at all.

In situations like this, it can be helpful to think in terms of sets as in Figure 11-11.

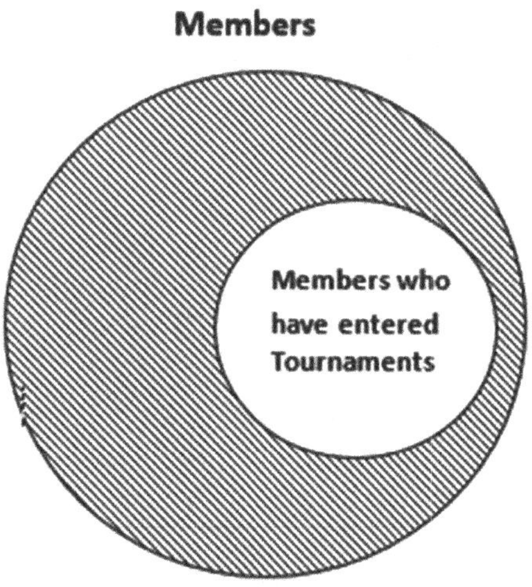

Members

Figure 11-11. *Finding members who have not entered tournaments by considering sets*

In Chapter 7 we looked at how to represent the difference between two sets by using the process approach and the keyword EXCEPT. The following query will return the IDs of members who have not entered a tournament:

```
SELECT MemberID FROM Member
EXCEPT
SELECT MemberID FROM Entry;
```

If we think in terms of the outcome approach we can describe the criteria for returning a particular MemberID. The following query is an example of using NOT IN to find the IDs of members who have never entered a tournament:

```
SELECT m.MemberID FROM Member m
WHERE m.MemberID NOT IN
    (SELECT e.MemberID FROM Entry e);
```

Chapter 7 has many examples of how to use nested queries such as this one.

All, Every

Wherever you see the words *all* or every in a description of a query you should immediately think of the division operator. Here are some examples of such queries:

- Find members who have entered *every* open tournament.

- Has anyone coached *all* the juniors?

Examples of the SQL to carry out these types of query are explained in detail in Chapter 7.

No Idea Where to Start?

Now let's look at the case where we have a good understanding of the intention of the natural language query and have an idea of which tables are involved. We've checked for some key words, but still feel confused. Now what? This is not uncommon (it happens to me regularly), so just relax.

When I have no idea where to start, I forget all about set operations and SQL. I stop thinking about tables, foreign keys, joins, and so on. Instead, I open the tables I think will be needed to answer the question and look at some of the data. I try to find examples that should be retrieved by the query. Then I try to write down the conditions that make that particular data acceptable.

This is the outcome approach describing *what* conditions the rows returned by the query should obey. It is a great way to proceed if you are having trouble deciding on the operations that could be involved in manipulating the tables (the process approach).

Let's try a query that stumped me a bit when I first thought of it: "Which teams have a coach as their manager?" The steps described here can really help.

Find Some Helpful Tables

Let's look at the key words in the query "Which teams have a coach as a manager?" We have the nouns "team," "coach," and "manager." We have a table called Team, and Coach and Manager are fields in the Member and Team tables, respectively. So the Team and Member tables look like a good place to start.

Try to Answer the Question by Hand

Next, take a look at the data in the tables and see how you would decide if a team had a coach as a manager. Figure 11-12 shows some relevant columns of the two tables. Can you find a team that satisfies the condition?

TeamName ▾	Manager ▾
TeamA	239
TeamB	153

Team table

MemberID ▾	LastName ▾	FirstName ▾	Coach ▾	Team ▾
118	McKenzie	Melissa	153	
138	Stone	Michael		
153	Nolan	Brenda		TeamB
176	Branch	Helen		
178	Beck	Sarah		
228	Burton	Sandra	153	
235	Cooper	William	153	TeamB
239	Spence	Thomas		
258	Olson	Barbara		
286	Pollard	Robert	235	TeamB
290	Sexton	Thomas	235	

Member table

Figure 11-12. How do we tell if a team has a coach as a manager?

We can find the IDs of the two team managers easily enough. They are the values in the Manager column of the Team table (239 and 153). Now, how do we check if these members are coaches? Looking at the Member table, we see that the coaches are in the Coach column. We need to check if either of our two managers appears in the Coach column. Member 153 does appear in the Coach column, so (TeamB) is managed by a coach.

Write Down a Description of the Retrieved Result

Figure 11-12 illustrates how we determined that TeamB has a coach as its manager. We now need to write a description of the logic that leads to that conclusion. This is where I like to use my fingers to point to the relevant rows to make it easier to describe the query, as in Figure 11-13.

TeamName ▾	Manager ▾
TeamA	239
t ☞ TeamB	153

Team table

MemberID ▾	LastName ▾	FirstName ▾	Coach ▾	Team ▾
118	McKenzie	Melissa	153	
138	Stone	Michael		
153	Nolan	Brenda		TeamB
176	Branch	Helen		
178	Beck	Sarah		
m ☞ 228	Burton	Sandra	153	
235	Cooper	William	153	TeamB
239	Spence	Thomas		
258	Olson	Barbara		
286	Pollard	Robert	235	TeamB
290	Sexton	Thomas	235	

Member table

Figure 11-13. *Naming the rows to help describe the required data*

We are going to check every team to decide if it should be retrieved. In Figure 11-13 this is represented by the finger labeled t, which will visit each row in turn. We can describe whether the current row meets the criteria as follows:

> *I'll write out the TeamName from row t in the Team table, if there exists a row m in the Member table where the value of coach m.Coach is the same as the manager of the team t.Manager.*

We can now translate this almost directly into SQL using a nested query (discussed in Chapter 4). One possible query would be:

```
SELECT t.TeamName FROM Team t
WHERE EXISTS
   (SELECT * FROM Member m WHERE m.Coach = t.Manager);
```

Are There Alternatives?

First attempts at queries aren't necessarily the most elegant. After all, we are following this route because we were stumped in the first place. This may not be a problem for the execution of the query, as the optimizer will likely find an efficient process. However, an inelegant SQL statement might be difficult for you and others to understand at a later time. Following the technique of solving the query by hand and describing the conditions often helps you understand what you are trying to do. That often makes the query seem much easier than you first thought.

Having made a first attempt at the query described in the previous section, we might realize that we could have thought of it this way: "The manager just has to be in the set of coaches." We can easily find the IDs of coaches with the query:

```
SELECT m.Coach FROM Member m;
```

We can then use that in a nested query, as shown here:

```
SELECT t.TeamName FROM Team t
WHERE t.Manager IN
    (SELECT m.Coach FROM Member m);
```

For me, the preceding query is simpler and easier to understand than the earlier one even though they have equivalent results.

We could have phrased the condition illustrated in Figure 11-13 like this:

> *If I have rows t in the Team table and m in the Member table then I'll write out the TeamName from row t in the Team table, if t.Manager = m.Coach*

Here is the preceding sentence translated into SQL:

```
SELECT t.TeamName FROM Team t, Member m
WHERE t.Manager = m.Coach;
```

The preceding query can be restated as a join:

```
SELECT t.TeamName
FROM Team t INNER JOIN Member m ON t.Manager = m.Coach;
```

Personally, I don't find the join particularly intuitive for this query. I doubt if someone else looking at the query would quickly understand its purpose.

Given there are several options for phrasing this query, it can be useful to check their relative efficiencies (as discussed in Chapter 10) if you think that might be important (unlikely in this case). If we add a DISTINCT phrase in the SELECT clause for the join queries then all four alternatives will produce the same result. For SQL Server 2012, each of the queries had the same execution plan, so they were all carried out in exactly the same way under the hood.

Checking Queries

We've written a query, run it, and retrieved some results. Is all well and good? Not necessarily. Just as first attempts at a query may not be elegant, neither may they be correct. Mistakes might arise from simple errors in the query syntax. These are usually easy to spot and correct. However, errors that result from subtle misunderstandings of the question or of the data can be more difficult to find.

I can't offer a foolproof way of checking that your query is correct, but I can give you some ideas for catching potential errors. Basically, they boil down to checking that you do not have extra, incorrect rows in your result and checking that you aren't missing any rows. In this section, we will look at ways to spot that your query might have a problem. In the next chapter, we will look at some of the common mistakes that might be behind the errors.

Check a Row That Should Be Returned

It is a good idea to have a rough idea of how many rows should be returned by your query: none, one, a few, or lots. If you get a surprising number then that can be a clue that something could be wrong. Next, take a look at your data and determine one record or row that should be returned by the query. In our example about teams with managers as coaches, we can check through the tables and find a team that satisfies the query. In Figure 11-13, we see that TeamB satisfies the conditions, so check that this team is in the output.

Remember that some queries may quite legitimately have no output. For example, it's perfectly reasonable that, with the data we have at any particular time, no teams are managed by a coach. However, your query must work in all situations. If it is at all possible, make a copy of the tables, alter the data so that a row meets the condition, and check that it is returned correctly.

Check a Row That Should Not Be Returned

Similar to checking for a row that should be returned, look through the data and find a team that *doesn't* have a coach as a manger. TeamA's manager (member 239) does not appear as a coach in the Member table, so make sure that team is *not* included in your output. Once again, it is a good idea to use some dummy data to check this if the real data does not cover all eventualities.

Check Boundary Conditions

If a query has any sort of numeric comparison, then as well as checking for example data that should be returned and that which shouldn't, we should also check the edge cases. Consider a query where we want to find people who have been members of our club for more than ten years. To be certain of the correctness, we need to check three possibilities:

- Make sure no record is returned for someone with less than 10 years of membership (for example, 8 years of membership).

- Make sure that someone who has belonged to the club for 12 years does get his record retrieved.

- Check for someone who has been a member for exactly 10 years.

The last boundary condition is always tricky. It comes down to an interpretation of the natural language question. Does "more than ten years" include people who joined in the season exactly ten years ago? Well, it probably does, given that a single season covers a whole year. With numerical comparisons of this sort the decision is whether we use > or >= in the select condition. It is important to check with users if there is any doubt about the intention of the query.

Finding data in the tables that are exactly on the boundaries is not always easy. However, it is usually possible to change the numeric value in your query to match the data. Find a particular member and change the value you are checking against in the query to match their years of membership. If Harry joined 16 years ago, change the query to compare with the value 16 and see if Harry is included (or not) as you expect.

Another important boundary condition, especially for aggregates and counts (covered in Chapter 8), is the value 0. Consider a query such as "Find members who have entered fewer than six tournaments." Doing a grouped by count on the Entry table will return some rows for sure, and we can check for those who have less than, more than, or exactly six entries. However, what about members who have never entered a tournament? They won't appear in the Entry table at all and will be missing from the results. So, whenever aggregates are involved, always check for what happens for a count of 0. For example, does your query return members who have entered *no* tournaments?

Check Null Values

Be aware that some of the values you are checking against may be nulls (discussed in Chapter 2). How does your query about team managers cope with the situation where the Manager field is null? Try it out on some dummy data and see. What do we expect (or want) to happen if there is a null in the JoinDate field when we run the query about length of membership?

Summary

The first rule about starting a query is to not panic. The next rule is to take small steps and look at the intermediate output to see if what you have done so far is helping you. Retain as many columns as possible in the initial queries so you can check that you understand what is happening.

Figure 11-14 gives a summary of some of the steps you can take when first starting out on a query. The diagram doesn't cover the whole process, but you should be able to make a reasonable start with these steps. Refer to the relevant chapters for more help.

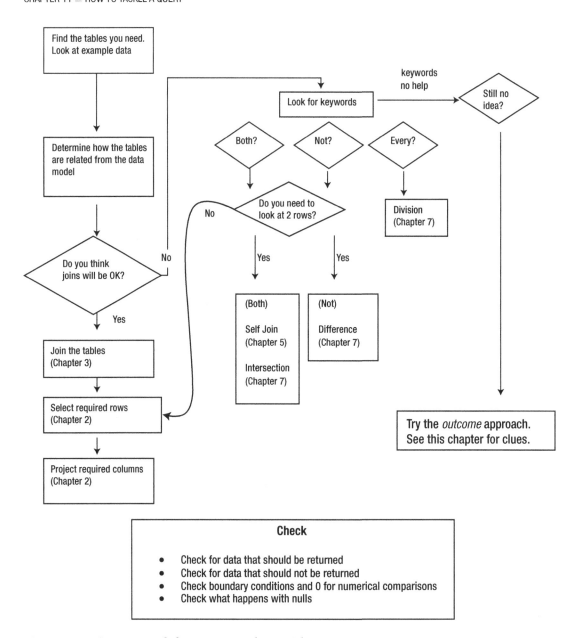

Figure 11-14. *Some steps to help you get started on a tricky query*

Common Problems

In this book, we've looked at different ways to tackle a variety of categories of queries. However, even if a query retrieves some valid-looking rows, all may not be well. In the previous chapter, we looked at the importance of checking the output to confirm that (at least some of) the expected rows are retrieved, as well as checking to make sure that (at least some) incorrect (or irrelevant) rows are *not* being returned.

The problems that can befall queries are not just a matter of having the wrong syntax in SQL statements, although that can certainly happen. Problems with the design of the tables or with data values can also affect the accuracy of queries. In this chapter, we will look at some common design and data problems, and also at some of the most common syntactic mistakes.

Poor Database Design

Good database design is absolutely essential to being able to extract accurate information. Unfortunately, you will sometimes be faced with databases that are poorly designed and maintained. Often there is not a great deal you can do. Sometimes you can extract something that looks like the required information, but it should be presented with a caution that the underlying data was probably inconsistent. In this section we look at some common problems and how they might be mitigated.

Data That Is Not Normalized

One of the most common data design mistakes is to have tables that are not normalized. We looked at an example of this in Chapter 1. Rather than having two tables, one for members and one for membership information such as fees, all this data was stored in a single table. As can be seen in Figure 12-1, this has the effect of storing the fee information several times.

© Clare Churcher 2016
C. Churcher, *Beginning SQL Queries*, DOI 10.1007/978-1-4842-1955-3_12

MemberID ▾	LastName ▾	FirstName ▾	Phone ▾	Handicap ▾	MemberType ▾	Fee ▾
118	McKenzie	Melissa	963270	30	Junior	$150.00
138	Stone	Michael	983223	30	Senior	$300.00
153	Nolan	Brenda	442649	11	Senior	$250.00
176	Branch	Helen	589419		Social	$50.00
178	Beck	Sarah	226596		Social	$50.00
228	Burton	Sandra	244493	26	Junior	$150.00
235	Cooper	William	722954	14	Senior	$300.00
239	Spence	Thomas	697720	10	Senior	$300.00
258	Olson	Barbara	370186	16	Senior	$300.00

Figure 12-1. *A non-normalized Member table containing fee information*

What happens now if we are asked to find the fee for senior members? The query here will result in two values: 300 and 250.

```
SELECT DISTINCT Fee
FROM Member
WHERE MemberType = 'Senior'
```

Although the two values retrieved by the query may be surprising, nothing is wrong with the query or the result. The value for Brenda Nolan, which is inconsistent with the other senior members, gives us the additional fee result. That value may be a typographical error, or it may indicate some sort of discount for Brenda, or it may be an instance of last year's fee that has not been updated. In either case, there is a problem with the design. The design should allow for regular fees for each grade to be recorded consistently and, if necessary, allow for storage of additional discounting regimes. At this point, other than redesigning the tables, there is nothing we can do but return the list of fees that have been recorded against the senior members. It is just worth understanding the underlying issues.

Another problem you may encounter is a single table that stores multivalued data. The versions of the club tables that we have been using allow a member to belong to just one team. The club may evolve to have several different types of teams (interclub teams, social teams, pairs, foursomes, and so on) that members can belong to at the same time. When the requirement for a second team to be recorded against a member arises, a common short-term fix is to add another Team column to the existing table. Figure 12-2 shows how the Member table might have evolved to allow members to be associated with up to three teams.

MemberID ▾	LastName ▾	FirstName ▾	Team1 ▾	Team2 ▾	Team3 ▾
118	McKenzie	Melissa			
138	Stone	Michael			
153	Nolan	Brenda	TeamB		
176	Branch	Helen	TeamA	TeamB	
178	Beck	Sarah			
228	Burton	Sandra	TeamC		
235	Cooper	William	TeamB		
239	Spence	Thomas	TeamA	TeamC	TeamB
258	Olson	Barbara			

Figure 12-2. *Poor table design to store more than one team for a member*

Now, suppose we are asked to find those members in TeamB. Brenda has TeamB in the Team1 column, Helen has TeamB in the Team2 column, and Thomas has TeamB in the Team3 column. We need to check every team column for the existence of TeamA. This isn't difficult, as the query here shows:

```
SELECT * FROM Member
WHERE Team1 = 'TeamB' OR Team2 = 'TeamB' OR Team3 = 'TeamB';
```

While we can extract the information we require from the table in Figure 12-2, the design is going to cause problems. We will have trouble if we have queries like "Find members who are in both TeamA and TeamB" or "Find members who are in more than two teams." You could probably devise queries that would answer these questions, but they would be ungainly. I would ask for the database to be redesigned properly before trying to fulfill such requests. If you meet resistance you can ask them what they will do if a member belongs to four teams or maybe twenty teams.

If members can belong to several teams we have a Many–Many relationship, which should be represented in a relational database with an intermediate Membership table[1] – something like the one in Figure 12-3.

MemberID ▾	Team
153	TeamB
176	TeamA
176	TeamB
228	TeamC
235	TeamB
239	TeamA
239	TeamB
239	TeamC

Figure 12-3. *A Membership table that records the relationship between members and teams*

The Membership table in Figure 12-3 records relationships between members and teams and is very similar to the Entry table, which records relationships between members and tournaments. The Membership table will need to be joined with the Member table to find the associated names, but if that is done we will have the same information as the one in Figure 12-2. With the new Membership table, we can now use all the relational operations, as described in previous chapters, to easily answer questions like "Who is in TeamA and TeamB?" and "Who is in three or more teams?"

We can create a Membership table with the following SQL code. The table includes only two foreign keys, to the existing Member and Team tables, and those fields also form a concatenated primary key.

```
CREATE TABLE Membership(
MemberID INT FOREIGN KEY REFERENCES Member,
Team CHAR(20) FOREIGN KEY REFERENCES Team,
PRIMARY KEY (MemberID, Team) );
```

---

[1]Refer to my book *Beginning Database Design* (New York: Apress, 20xx) for more information.

If you don't mind a bit of manual fiddling about, you can populate the new Membership table with repeated update queries like the one here:

```
INSERT INTO Membership (MemberID, Team)
SELECT MemberID, 'TeamA'
FROM Member
WHERE Team1 = 'TeamA' OR Team2 = 'TeamA' OR Team3 = 'TeamA'
```

The query finds each member who is in TeamA and creates an appropriate row in the Membership table. If there are not too many teams, you can manually alter the second and last lines of the query for each team (TeamA, TeamB, and so on) and populate the new Membership table quite quickly. You then need to delete the Team columns from the Member table in Figure 12-2, and the database will be greatly improved.

Tables with No Primary Key

The previous section gave an example of the problems you can run into if the underlying database has inappropriate tables. You will sometimes find that the database has the correct tables, but they do not have suitable primary or foreign key constraints. In these cases, the underlying data values are likely to be inconsistent. While your queries may be correctly formed, the results will be unreliable. In this section, you will see how you can use queries to find some inconsistencies that may be present in your data.

Suppose that the Membership table in Figure 12-3 had been created without a primary key. This would allow the table to have duplicate rows. For example, we might have two identical rows for member 153 being on TeamB.[2] A query to count the number of members on TeamB will produce an incorrect result.

If you try to add a primary key when duplicates already exist, you will get an error. This is one way to find where problems are! Before you can add a primary key you will need to find the duplicated rows and investigate how to resolve the issue. One convenient way to find duplicated values is to do a GROUP BY query (see Chapter 7) on the fields that should be unique and use a HAVING clause to find those with two or more entries. The following query will return duplicated values for our potential primary key fields MemberID and Team:

```
SELECT MemberID, Team, Count(*)
FROM Membership
GROUP BY MemberID, Team
HAVING Count(*) > 1;
```

If the table has fields other than the primary key fields, you need to manually inspect the values in those columns to decide which row should be deleted. The Membership table, which has only primary key fields, causes a different problem. How do we delete just one copy of the row for member 153 in TeamB? Because the entire rows are the same, we can't differentiate them, and so any query that deletes one row will delete both. You software might have a tabular-like interface that will allow you to delete just one of the rows, but if not you may have to delete both rows and manually add one back. If there are a lot of duplicate values, then another way to resolve the situation is to create a new table and then insert just the distinct values from the original table. The following query shows how to populate the new table NewMembership:

```
INSERT INTO NewMembership
SELECT DISTINCT MemberID, Team
FROM Membership;
```

---

[2]This is the difference between a relation that is defined as having unique tuples and a table that can have duplicate rows. See Appendix 2 for further information.

You will then need to remove all the foreign key constraints referencing the old table, delete that table, rename the new table, and recreate the foreign keys. It's easier to make sure every table has a primary key from the start!

Tables with Missing Foreign Keys

Another problem is having a Membership table (as in Figure 12-3) with no foreign key constraints. We can then find ourselves with the problem of having a row for member 1118 being in TeamA when no member 1118 is listed in the Member table. We will not be able to add a foreign key constraint if the data has this sort of problem.

There are several ways to find such values of MemberID in the Membership table that do not have a matching entry in the Member table. One way is to use a nested query (discussed in Chapter 4), as shown here:

```
SELECT ms.MemberID FROM Membership ms
WHERE ms.MemberID NOT IN
    (SELECT m.MemberID FROM Member m);
```

Having found the unmatched values for MemberID, we will then have to decide if it is a typographical error or if we are missing a member from the Member table.

When the data is in a consistent state it will be possible to add a foreign key constraint to the Membership table to make sure it stays that way. The following query will add the constraint to the MemberID field:

```
ALTER TABLE Membership
ADD FOREIGN KEY (MemberID)
REFERENCES Member;
```

Similar Data in Two Tables

Sometimes a database might have extra tables that are not required and will cause problems. An example for our club database might be having a separate table for coaches or managers, as shown in Figure 12-4. The rationale might have been that the extra table would make it easier to create lists of coaches and their phone numbers (which would otherwise require a self join or nested query).

MemberID ▾	LastName ▾	FirstName ▾	Phone ▾
153	Nolan	Brenda	442649
235	Cooper	William	685563

Coach Table

MemberID ▾	LastName ▾	FirstName ▾	Phone ▾	Handicap ▾	MemberType ▾	JoinDate ▾	Coach ▾	Gender ▾
118	McKenzie	Melissa	963270	30	Junior	28-May-05	153	F
138	Stone	Michael	983223	30	Senior	31-May-09		M
153	Nolan	Brenda	442649	11	Senior	12-Aug-06		F
176	Branch	Helen	589419		Social	06-Dec-11		F
178	Beck	Sarah	226596		Social	24-Jan-10		F
228	Burton	Sandra	244493	26	Junior	09-Jul-13	153	F
235	Cooper	William	722954	14	Senior	05-Mar-08	153	M
239	Spence	Thomas	697720	10	Senior	22-Jun-06		M
258	Olson	Barbara	370186	16	Senior	29-Jul-13		F
286	Pollard	Robert	617681	19	Junior	13-Aug-13	235	M
290	Sexton	Thomas	268936	26	Senior	28-Jul-08	235	M

Member Table

Figure 12-4. *An additional table for coaches can lead to inconsistent data*

The additional table will inevitably cause problems. In Figure 12-4, we already see inconsistent data for William Cooper's phone number. The only real cure is to get rid of the extra table.

If the purpose of an additional table like the one in Figure 12-4 is unclear, we can use set operations to investigate which members appear in each of the tables. The intersection operator will find rows for people who are in both tables, and the difference operator will find those people who are in one and not the other. This may help with understanding what the tables represent.

Once the design is correct, creating a view that shows the coach information would be helpful for users who don't want to be creating self joins every time they want information just about coaches. The following query does the trick:

```
CREATE VIEW CoachInfo AS
SELECT * FROM Member
WHERE MemberID IN
    (SELECT Coach FROM Member);
```

Inappropriate Types

Having the fields in a table created with inappropriate types is another problem that can make queries look as though they are not behaving. I've seen whole databases where every field is a default text field.

Having the wrong field type means the data misses a great deal of validity checking. For example, if our Member table had all text fields, we could end up with values like "16a" or "1o" in the Handicap column, which should only have integer numbers, or text like "Brenda" in the Coach column, which should only contain IDs of members.

Incorrectly entered values aside, inappropriate types give rise to other problems. Each type has its own rules for ordering values. Text types order alphabetically, numbers order numerically, and dates order chronologically. Different orderings clearly will be an issue if we add an ORDER BY clause to a query. A text field containing numbers will order alphabetically, giving an order like "1," "15," "109," "20," "245," and "33," as described in Chapter 2.

Incorrect types also cause problems when making comparisons. If we ask for values to be compared, the comparison used will depend on how the particular field type involved is ordered. For numbers entered in a text field, we will get comparisons such as "109" < "15" or "33" > "245" as per the ordering described in the previous paragraph. This will cause some odd output if we ask for people with handicaps less than 5, for example. It can be difficult to sort out what is going wrong, because the query syntax is fine and the data appears to be OK. Going behind the scenes to check out the data type might not be something that is immediately obvious.

It is possible to change the type of a column in an existing table, but I find it a bit scary. For example, if you change from text to numeric values, "10" will probably be fine but "1o" will cause an error. I prefer a more conservative approach: I make a new table with the appropriate types, and then insert the old values with the aid of a conversion function. The query that follows shows how we could populate a new table NewMember with IDs and names and with the old text values for the Handicap column converted to numeric values:

```
INSERT INTO NewMember (MemberID, LastName, FirstName, Handicap)
SELECT MemberID, LastName, FirstName, CONVERT(INT Handicap)
FROM Member;
```

This way, we still have the original data if the conversions result in something unexpected.

Problems with Data Values

Even with a well-designed database, we still have the issue of the accuracy of the data that has been entered. As the query designer, you can't be held responsible for some accuracy problems. If a person's address has been entered incorrectly, there is not much anyone can do to find or fix the problem (apart from waiting for the mail to be returned to sender). However, you can be aware of a number of things, and even if you can't fix the problems, you can at least raise some alarms. In addition, it is sometimes possible to fix some problem data with careful use of update queries.

Unexpected Nulls

Nulls can cause all sorts of grief in databases. The real problem (as discussed in Chapter 2) is that a null can mean either that the value is unknown or that the value doesn't apply for a particular record. If a member in our club has a null value for his Team field, it could mean he isn't on a team or it could mean that he is on a team but we haven't recorded which one. As with other data problems, there is not much we can do about this. However, with something like the Gender field, we know that for the golf club, all members need to identify as either male or female. The nulls mean that for some members the gender has not been recorded. The same applies to fields like date of birth.

If, for example, you are asked for a list of the men in the club, it is often a good idea to also run another query for those rows where Gender IS Null. You can then say to your client, "Here are the men, and here are the members I'm not sure about." Such an approach can help avoid letters from aggrieved gentlemen who don't appear on the list.

Be aware of the differences between queries with the following two counts: COUNT(*) and COUNT(Gender). The first will count all the rows in the database; the second will count all the rows with a non-null value for gender. In the ideal golf club, these would be the same. In practice, they may not be.

Incorrect or Inconsistent Spelling

Any database will have spelling mistakes in the data at some point. Mr. Philips may appear as Phillips, Philipps, or Philps for various reasons, ranging from illegible handwriting on the application form to a simple data-entry mistake. If you are trying to find information about Mr. Philips and you suspect there might be a problem, you can use functions or wildcards to find similar data. Different products have different ways of doing this.

We can use the keyword LIKE to find similar spellings. The wildcard symbol % (* in Access) stands for any group of characters. Our several versions of spelling for Philips would all be retrieved by the following query:

```
SELECT * FROM Member
WHERE LastName LIKE 'Phil%';
```

Another problem involving incorrect or inconsistent spelling arises when you might be expecting a particular set of values or categories in a field. For example, in our Member table, we might be expecting values M or F in the Gender column, but there may be the odd male or m value. In the MemberType column, we expect Junior, Senior, or Associate, but in practice may find jnior or senor. If the tables have been designed with appropriate check constraints or foreign keys, this won't be a problem. However, often these constraints are not present, so it is useful to check for problematic entries with a query such as the one here:

```
SELECT * FROM Member
WHERE MemberType NOT IN ('Senior', 'Junior', 'Associate');
```

Having found the rows that do not conform to expectations it may be possible to amend the data and then apply a check constraint so that it remains consistent. For example, the following query will apply a constraint on the MemberType field so that only the valid values can be entered:

```
ALTER TABLE Member
ADD CONSTRAINT Chk_type CHECK(MemberType IN
    ('Senior', 'Junior', 'Associate'));
```

Extraneous Characters in Text Fields

A common problem when trying to retrieve data that matches a text value is leading or trailing spaces and other nonprintable characters that have found their way into the data.

If we have a field like FirstName in our database, for example, we may find that there are some spaces before or after the name. Sometimes, if a character field is specified as being a particular length, trailing spaces may be added. If a row has a name has been stored as ' Dan ' then a WHERE clause with the condition FirstName = 'Dan' may not retrieve that row. Most database software will have several functions for dealing with text. There are likely to be forms of *trim* functions, which remove spaces from the start and end of text values. Check out your documentation to see what your implementation has.

The RTRIM() function in the SQL statement that follows will strip any spaces from the right end of the FirstName value before making the comparison:

```
SELECT * FROM Member
WHERE RTRIM(FirstName) = 'Dan';
```

The preceding query does not strip the spaces from the field permanently. The RTRIM() function just returns a value without the spaces in order to make the comparison. However, you can use update queries to permanently remedy some of these data inconsistencies. The query that follows shows how to ensure no values in the FirstName column of the Member table have any leading (LTRIM()) or trailing (RTRIM()) spaces. It essentially replaces all the values with trimmed values:

```
UPDATE Member
SET FirstName = RTRIM(LTRIM(FirstName));
```

A more disturbing problem is characters that look like spaces but are actually some other white space characters. This sometimes occurs when data is cut and pasted or otherwise moved between various products and different implementations. This can take some tracking down.

Two other data-entry gotchas are the numbers 0 (zero) and 1 (one) being entered instead of the letters o (oh) and l (el). You can spend hours trying to debug a query that is looking for "John" or "Bill," but if the underlying data has been mistakenly entered as "J0hn" or "Bi11" you will search in vain.

The moral is that weird things can happen with data values, so when the troubleshooting of your query syntax fails, check the underlying data.

Inconsistent Case in Text Fields

If your SQL implementation is case sensitive, you need to be aware that some data values may not have the expected case. Dan may have had his first name incorrectly entered into the Member table as "dan." In case-sensitive implementations, a query with the clause WHERE FirstName = 'Dan' will not retrieve his information. As mentioned in Chapter 2, using a function that converts strings of characters to uppercase will help find the right rows. In the query that follows we convert FirstName (temporarily) to uppercase, and then compare that with the uppercase rendition of what we are seeking:

```
SELECT * FROM Member
WHERE UPPER(FirstName) = 'DAN';
```

It is quite difficult to find problems with case in names because not all names conform to being lowercase with an uppercase first letter; for example, de Vere and McLennan. But, for fields like Gender (M or F) or MemberType (Junior, Senior, or Associate), we know what we expect the values to be. The best way to ensure that they are consistent is to put a check constraint on the field as discussed earlier in this chapter.

Diagnosing Problems

In the previous sections, we saw problems that can arise with poor database design and inconsistent or incorrect data. Much of the time, however, if the result of your query is not looking quite right, it is probably because you have the wrong SQL statement. The statement may be retrieving rows that are different from what was expected. In Chapter 10 there is a section on some ways that you can check to see if the result of a query is what is expected.

In the previous chapter, I suggested a way to approach queries that lets you build the query up slowly so you can check that each step is returning appropriate rows. However, if you are presented with a full-blown, complex query that is not delivering as expected, you need to pare it down until you find where the problem lies. If you have noticed a problem, then you have a good place to start. You have either noticed an expected row is missing or that a row not satisfying the requirements has been retrieved. Concentrate on finding where in the query that problem is. The following sections offer some suggestions.

Check Parts of Nested Queries Independently

Where you have one query nested inside another, the first thing to check is that the nested part is behaving itself. Take a look at this query:

```
SELECT *
FROM Member m
WHERE m.MemberType = 'Junior' AND Handicap <
      (SELECT AVG(Handicap)
      FROM Member);
```

If you are having trouble with a query like this, cut and paste the inner query and run it independently. Check to see if it is returning the correct result. If this is OK, you can try doing the outer query on its own. To do this, just put some value in place of the inner query (such as Handicap < 10) and see if that returns the correct results. If you can narrow down the problem to one part of the query, you have made a good start.

This approach doesn't work if the inner and outer parts of the query are related (see Chapter 4), but some of the following techniques might help with that situation.

Understand How the Tables Are Being Combined

Many queries involve combining tables with relational operations (join, union, and so on). Make sure you understand how the tables are being combined and whether that is appropriate. Consider a query such as the following:

```
SELECT m.LastName, m.FirstName
FROM Member m, Entry e, Tournament t
WHERE m.MemberID = e.MemberID
AND e.TourID = t.TourID AND t.TourType = 'Open' AND e.Year = 2014;
```

Three tables are involved in this query. It might take a moment to figure out that they are being joined. Make sure that is appropriate for the question being asked. Chapter 10 has examples of keywords in questions and the appropriate ways to combine tables.

Remove Extra WHERE Clauses

After combining tables, usually only some of the resulting rows are required. In the query in the previous section, only part of the WHERE clause is needed for the join operations. After the join, only the rows satisfying t.TourType = 'Open' AND e.Year = 2014 are retained. If you have rows missing from your result, it is often useful to remove the parts of the WHERE clause that are selecting a final subset of the rows after the join. If the rows are still missing, then you know that (for this example) the problem is occurring in the join.

Retain All the Columns

I'm a big fan of always saying SELECT * in the early stages of developing queries that involve joins. If we suspect a problem with the joins, then by leaving all the columns visible, we can see if the join conditions are behaving as expected. Once we are happy with the rows being retrieved, we can retain just the columns required.

However, if we are combining tables with set operations, this approach will be counterproductive, as projecting the right columns is critical (see the "Do You Have Correct Columns in Set Operations" section later in this chapter).

Check Underlying Queries in Aggregates

If you have a problem with a query involving an aggregate (for example, `SELECT AVG(Handicap) FROM ...` `WHERE ...`) check that you have retrieved the correct rows before the aggregate function is applied. Change the query to `SELECT * FROM ... WHERE ...`, and confirm that this returns the rows for which you want to find the average. In fact, I recommend always doing this with an aggregate, because it is difficult to otherwise check if the numbers being returned are correct.

Common Symptoms

Having tried some of the steps in the previous chapter, you will have simplified your query to isolate where the problem is. In this section, we will look at some specific symptoms and some likely causes.

No Rows Are Returned

It is usually easy to spot a problem with your query when no rows are returned and you know that some should be. Questions that involve "and" or "both" can often have this problem. For example, consider a question such as "Which members have entered tournaments 24 and 36?" A common first attempt (and I still catch myself doing this sometimes) is a query statement such as:

```
SELECT * FROM Entry
WHERE TourID = 24 AND TourID = 36;
```

The preceding query asks for a row from the `Entry` table where `TourID` simultaneously has two different values. This never happens, and so no rows are retrieved. The cure is to use a self join (covered in Chapter 5) or an intersection operation (covered in Chapter 7).

Getting no rows returned from a query may also be an extreme example of one of the problems in the next section.

Rows Are Missing

It can be difficult to spot if some rows are being missed by your query, especially when the set of retrieved rows is large. If you get 1,000 rows returned, you might not notice that one is missing. Careful testing is required, and some ideas for how to do this were discussed in Chapter 10. It is often worthwhile to run through the following list of common errors to see if any might apply.

Should You Have an Outer Join?

Using an inner join when an outer join is required is a very common problem. Suppose that we are trying to get a list of member information that includes names and fees. For this, we need the `Member` table (for the names) and the `Type` table (for the fees). A first attempt at a query might be as follows:

```
SELECT m.LastName, m.FirstName, t.Fee
FROM Member m, Type t
WHERE m.MemberType = t.Type;
```

We know there are, say, 135 members, but we are getting only 133 rows from the query. The issue here is that we are performing an inner join (see Chapter 3), so any members with a null value for member type will not appear in the result. Of course, this may be the result you want (those members who have a type and fee), but it is not the correct output if you want a list of all members and the fees for those who have them.

An outer join (also discussed in Chapter 3) that includes all the rows of the Member table will solve this problem. Whenever you have a join, it is worth thinking about the join fields and considering what you want to happen where a row has a null value in that field.

Have Selection Conditions Dealt with Nulls Appropriately?

Nulls can cause quite a few headaches if you forget to consider their effect on your queries. The previous section looked at nulls in a joining field. You also need to remember to check for comparisons involving fields that may contain nulls. We looked at this in Chapter 2 and also earlier in this chapter.

Consider two queries on the Member table with selection conditions Gender = 'M' and Gender <> 'M'. It is reasonable to think that all rows in the Member table should be returned by one of these queries. However, rows with a null in the Gender field will return false for both these conditions (any comparison with a null returns false), and the row will not appear in either result.

Say we want to get a list of members of our club who are not particularly good players (to offer them coaching, perhaps). Someone may suggest a query like the following to find members who do not have a low handicap:

```
SELECT *
FROM Member m
WHERE m.Handicap > 10;
```

The problem is that the preceding query will miss all the members with no handicap. Altering the WHERE condition to m.Handicap > 10 OR m.Handicap IS Null will help in this situation.

Are You Looking for a Match with a Text Value?

It is very disturbing to be trying to find rows for Jim, to be able to see Jim in the table, and to have your query return nothing. This may be caused by one of the problems we looked at in the "Problems with Data Values" section earlier in this chapter.

One quick way to eliminate the possibility of dodgy text values is to use LIKE for comparisons. For example, where you have = 'Jim', replace it with LIKE '%Jim%'. If the query then finds the row you were expecting (possibly along with some others), you know the problem is with the data. As noted earlier, putting the wildcard % (or * in Access) at the beginning and end of the string will find leading or trailing spaces and other nonprintable characters.

Have You Used AND Instead of OR?

We discussed the problem of queries involving the words *and* or *or* in the previous chapter (in the "Spotting Key Words in Questions" section). I'll recap briefly. The word *and* can be used in natural English to describe both a union and an intersection. When we say "women *and* children," we usually mean the *union* of the set of females and the set of young people. When we say "cars that are small *and* red," we mean the *intersection* of the set of small cars and the set of red cars.

If we are looking for "women and children" and use the selection condition `Gender = 'F' AND age < 12`, we are actually retrieving the intersection of women and children (or girls). We need the condition to be `Gender = 'F' OR age < 12`.

It is very easy to unwittingly translate the *and* in the English question to an `AND` in the query inappropriately, which can result in missing rows. If in doubt, try drawing the Venn diagrams described in the previous chapter.

Do You Have Correct Columns in Set Operations?

If your query involves intersection or difference operations, the result may have fewer rows than expected because you have projected the wrong columns initially. We looked at this in Chapter 7. Here is a brief example for intersection; the same issue applies to difference operations as well.

We want to find out who has entered both tournaments 25 and 36. We realize that we need an intersection and try the following query:

```
SELECT * FROM Entry
WHERE TourID = 25
INTERSECT
SELECT * FROM Entry
WHERE TourID = 36;
```

No rows will be returned from this query, regardless of the underlying data. The intersection finds rows that are exactly the same in each set. However, all the rows in the first set will have 25 as the value for `TourID` 25, and all the rows in the second set will have the value 36. There can never be a row that is in both sets. What we are looking for is the member IDs that are in both sets, so the `SELECT` clauses in each part of the query should be `SELECT MemberID FROM Entry`.

The preceding query is an extreme example of retaining the wrong columns, resulting in no rows being returned. The discussion around Figure 7-14 in Chapter 7 shows how retaining different columns in intersection and difference queries can result in very different results. You need to ensure that you are retaining the columns that are appropriate for the question being asked.

More Rows Than There Should Be

It is often easier to spot extra rows than it is to notice that rows are missing from your query result. You only need to see one record that you weren't expecting, and you can concentrate on the different parts of your query to see where it failed to be excluded. Here are a couple of causes of extra rows.

Did You Use NOT Instead of Difference?

With questions containing the words *not* or *never*, a sure way to get extra rows is to use a condition in a `WHERE` clause when you really need a difference operator. We looked at this issue in Chapter 4. To recap, consider a question like "Which members have never entered tournament 25?" A common first attempt using a select condition is:

```
SELECT * FROM Entry
WHERE TourID <> 25;
```

The condition in the WHERE clause checks rows one at a time to see if they should be included in the result. If there is a row for member 415 entering tournament 36, then that row will be retrieved, regardless of the possibility that another row shows member 415 entered tournament 25. For example, if member 415 has entered tournament 25 and four other tournaments, we will retrieve four rows when we were expecting none.

The correct procedure for this type of question is to use a nested query (see Chapter 4) or the EXCEPT difference operator (see Chapter 7). We need to find the set of all members (from the Member table) and remove the set of members who have entered tournament 25 (from the Entry table).

If we employ the process approach we might come up with the following query, which looks for the difference between the two sets:

```
SELECT MemberID FROM Member
EXCEPT
SELECT MemberID FROM Entry
WHERE TourID = 25;
```

If we started with an outcome approach we might have arrived at a nested query, as here:

```
SELECT MemberID FROM Member
WHERE MemberID NOT IN
    (SELECT MemberID FROM Entry
     WHERE TourID = 25);
```

Have You Dealt with Duplicates Appropriately?

It sometimes takes a little thought to decide what needs to be done with duplicate records retrieved from a query. By default, SQL will retain all duplicates. The following two requests sound similar:

- Give me a list of the names of my customers.

- Give me a list of the cities my customers live in.

In the first, we probably expect as many rows as we have customers; if we have several Johns, we expect them all to be retained. In the second, we expect one row per city. If we have 500 customers living in Christchurch, we don't expect all 500 rows to be returned.

In the query to find the cities, we want only the distinct values, so we should use the DISTINCT keyword:

```
SELECT DISTINCT (City) FROM Customer;
```

Incorrect Statistics or Aggregates

If we are using aggregates such as counting, grouping, or averaging and the underlying query misses rows or returns extra rows, then clearly the statistics will be affected. A couple of other things to consider are how nulls and duplicates are being handled.

SQL will not include any null fields in its statistics. For example, COUNT(Handicap) or AVG(Handicap) will ignore any rows with nulls in the Handicap field. It is also important to consider what you want done with duplicates, especially for counting functions. COUNT(Handicap) will return the number of members who have a value in the Handicap column. COUNT(DISTINCT Handicap) will return the number of different values in the Handicap column; if all the members have a handicap of 20, it will return a count of 1.

The Order Is Wrong

If you have used an ORDER BY clause in your query and you are having problems with the order in which the rows are being presented, there is often a problem with the underlying data. Review the "Problems with Data Values" section earlier in this chapter. Check that the field types are appropriate (for example, numeric values aren't being stored in text fields) and that text values have consistent case and no extraneous characters.

Common Typos and Syntax Problems

Sometimes a query doesn't run because of some simple problem with the syntax – that is, the way the query is worded. Syntax problems involve things like missing parentheses or incorrect spellings of fields or keywords. Hopefully the database software will alert you if there is a problem with the syntax, but, as some editors are quite basic, that may or may not be helpful in finding and correcting the problem. Here are a few things to check:

- *Quotation marks*: Most versions of SQL require single quotation marks around text values, such as 'Smith' or 'Junior', although some use double quotation marks in some circumstances. If you are cutting and pasting queries, be sure the correct quotation marks have been transferred. When I cut and paste the queries in this book from Word to Access, the quotation marks look OK, but I need to re-enter them. Also check that all the quotation marks are paired correctly. Don't use quotes around numeric values. Something like Handicap < '12' will cause problems if Handicap is a numeric field.

- *Parentheses:* These are required in nested queries and also can be used to help readability in many queries (such as those with several joins). Check that all the brackets are paired correctly.

- *Names of tables and fields*: It seems obvious that you need to get the names of tables and fields correct. However, sometimes a simple misspelling of a table name or field can cause an unintelligible error message. Check carefully.

- *Use of aliases*: If you use an alias for table names (for example, Member m), check that you have associated the correct alias with each field name.

- *Spelling of keywords*: Some software for constructing SQL queries will highlight keywords, so it is very apparent if you have spelled them incorrectly. If your version doesn't show this, then check keyword spelling, too. I often type FORM instead of FROM or AVERAGE() instead of AVG().

- IS Null *versus* = Null: Some versions of SQL treat these quite differently. IS Null always works if you are trying to find fields with a null value.

Summary

Before you can correct a query, you need to notice that it is wrong in the first place. It is preferable that we find potential problems before our users find them for us. Always check the rows returned from a query, as described in the previous chapter. When you do discover errors, the following are some ideas for tracking down the cause of the problem:

- Check that the underlying tables are combined appropriately (join, intersection, and so on).

- Simplify the query by removing selection conditions and aggregates to ensure the underlying rows are correct.

- Retain all the columns in a query with joins until you are sure that the tables have been combined appropriately.

- Check each part of nested queries or queries involving set operations independently.

- Check queries for questions with the words *and* or *not* to ensure you have not used selection conditions when you need a set operation or nested query.

- Check that the columns retained in queries with set operations are appropriate.

- Check that nulls and duplicates have been dealt with properly.

- Check that underlying data types are correct and that data values are consistent.

APPENDIX 1

Example Database

Most of the examples in this book use the golf club database. Visit the catalog page for this book on the Apress website, look under the Source Code/Downloads tab, and you will find an Access version of this database and also the SQL scripts for creating and populating the tables. Figure A1-1 shows how the tables in the database are related, and Figure A1-2 shows the data in the tables.

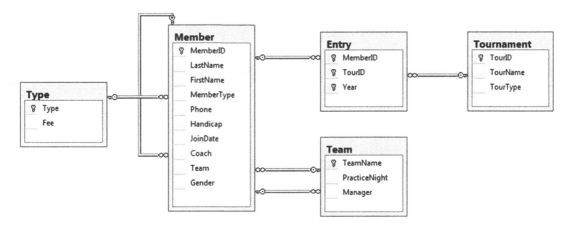

Figure A1-1. *The data model for the golf club database*

© Clare Churcher 2016
C. Churcher, *Beginning SQL Queries*, DOI 10.1007/978-1-4842-1955-3_13

MemberID	LastName	FirstName	Handicap	Gender	Team	MemberType	Coach	Phone	JoinDate
118	McKenzie	Melissa	30	F		Junior	153	963270	28-May-05
138	Stone	Michael	30	M		Senior		983223	31-May-09
153	Nolan	Brenda	11	F	TeamB	Senior		442649	12-Aug-06
176	Branch	Helen		F		Social		589419	06-Dec-11
178	Beck	Sarah		F		Social		226596	24-Jan-10
228	Burton	Sandra	26	F		Junior	153	244493	09-Jul-13
235	Cooper	William	14	M	TeamB	Senior	153	722954	05-Mar-08
239	Spence	Thomas	10	M		Senior		697720	22-Jun-06
258	Olson	Barbara	16	F		Senior		370186	29-Jul-13
286	Pollard	Robert	19	M	TeamB	Junior	235	617681	13-Aug-13
290	Sexton	Thomas	26	M		Senior	235	268936	28-Jul-08
323	Wilcox	Daniel	3	M	TeamA	Senior		665393	18-May-09
331	Schmidt	Thomas	25	M		Senior	153	867492	07-Apr-09
332	Bridges	Deborah	12	F		Senior	235	279087	23-Mar-07
339	Young	Betty	21	F	TeamB	Senior		507813	17-Apr-09
414	Gilmore	Jane	5	F	TeamA	Junior	153	459558	30-May-07
415	Taylor	William	7	M	TeamA	Senior	235	137353	27-Nov-07
461	Reed	Robert	3	M	TeamA	Senior	235	994664	05-Aug-05
469	Willis	Carolyn	29	F		Junior		688378	14-Jan-11
487	Kent	Susan		F		Social		707217	07-Oct-10

Member Table

MemberID	TourID	Year
118	24	2014
228	24	2015
228	25	2015
228	36	2015
235	38	2013
235	38	2015
235	40	2014
235	40	2015
239	25	2015
239	40	2013
258	24	2014
258	38	2014
286	24	2013
286	24	2014
286	24	2015
415	24	2015
415	25	2013
415	36	2014
415	36	2015
415	38	2013
415	38	2015
415	40	2013
415	40	2014
415	40	2015

Entry Table

TeamName	PracticeNight	Manager
TeamA	Tuesday	239
TeamB	Monday	153

Team Table

TourID	TourName	TourType
24	Leeston	Social
25	Kaiapoi	Social
36	WestCoast	Open
38	Canterbury	Open
40	Otago	Open

Tournament Table

Type	Fee
Associate	60
Junior	150
Senior	300
Social	50

Type Table

Figure A1-2. *The tables and data for the golf club database*

APPENDIX 2

■ ■ ■

Relational Notation

Relational database theory is based on set theory.[1] When we query a database we are essentially formulating a question to retrieve a subset containing the information we require. There are two approaches for retrieving a subset of data. *Relational algebra* is a description of the operations to perform on the data (in the body of the book we called this the process approach). *Relational calculus* describes conditions that the retrieved data must satisfy (we referred to this as the outcome approach). In this appendix we introduce the formal notation to formulate queries using relational algebra and calculus. This will allow you to think about queries from a different perspective. If you are interested in following up on the formal mathematics, There are more theoretical publications available.[2] No new concepts are presented here that have not been discussed previously–it is just the notation that is different. The more formal notation allows queries to be expressed very concisely, and the underlying mathematics can be useful when dealing with complex situations. We will use the database described in Appendix 1 for the examples.

Introduction

As an example of how thinking of data as sets can help us, let's consider a set that contains information about all the people on Earth. We can define a subset that contains all the men, another that contains all the golfers, another that contains people over 40, and another that contains Italians. These sets can all overlap, as shown in the diagram in Figure A2-1. This type of diagram is called a Venn diagram.

---

[1]The relational theory was first introduced by the mathematician E. F. Codd in June 1970 in his article "A Relational Model of Data for Large Shared Data Banks" in *Communications of the ACM*: 13, pp. 377–387.
[2]For example: *Databases in Depth: Relational Theory for Practitioners* by C.J. Date (City, state: O'Reilly, 2005).

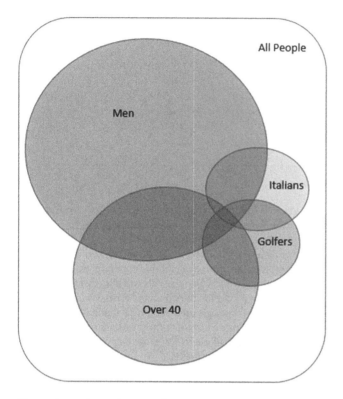

Figure A2-1. *Venn diagram showing subsets of people*

Figure A2-1 helps us visualize the sets that satisfy criteria such as *Italian men over 40 who play golf* (the area where all the circles overlap) or *people who don't play golf* (everywhere in the large rectangle except the Golfers circle). These two areas are easy to describe; however, it is not always simple to define the subset we require. The area containing *Italian golfers who are 40 or under* takes a bit more effort to find, and it is difficult to describe without the diagram to help.

A database is only useful if you can accurately extract the appropriate subset of data when you need it. As the criteria become more complex and the number of tables increases, it can become difficult to keep everything in your head and correctly describe what you are trying to find. It is in these more complex situations that having a more formal and succinct notation can be very helpful.

Relations, Tuples, and Attributes

It is common to think of a database as a number of tables. A table (e.g., `Person`) will have a several columns. Each row in the table represents an individual person with the appropriate values for that person appearing in each column. More formally, a database is referred to as a *set of relations*, and each relation is a *set of tuples*. A *tuple* is a set of *attribute values*; for example, {Ali, Brown, 2/8/1967}.

A relation consists of a *heading* and a *body*. The heading is a description of the data that is contained in the relation. Part of that description is a set of attribute names; for example, {`FirstName`, `LastName`, `Date_of_Birth`}. In addition, each attribute has a *domain*, or set of allowed values. For example, `Date_of_Birth` must be a valid date. A domain can be a primitive type (e.g., integer, string) or a user-defined type (e.g., `WeekDays` = {Mon, Tue, Wed, Thu, Fri, Sat, Sun}). A database *schema* is the set of headings for all the relations plus any constraints that have been defined.

The body of the relation contains the data values. It consists of tuples containing values for each of the attributes.

Table A2-1 shows analogous terms for the two ways of describing a database.

Table A2-1. *Comparative Terms*

Relational Term	Database Term
Database (set of relations)	Database
Relation (set of tuples)	Table
Tuple (set of attribute values)	Row
Attribute name	Column Name
Domain	Column datatype (primitive or user defined)

The main differences between an (unkeyed) table and a relation or set of tuples are that there is no order to the tuples and each tuple must be unique.

Figure A2-2 shows how we can visualize a relation as a set of tuples.

Figure A2-2. *A relation is a set of tuples.*

The traditional way of representing a set in a Venn diagram, as in Figure A2-1, reinforces the concept that there is no order to elements in a set. There is no first or next or previous element. The usual format for a table can imply that the rows have some sort of intrinsic order. When you query a database then, theoretically, the tuples returned have no guaranteed order unless you specify an order as part of the query. In practice, a simple query is likely to return rows in the same order each time it is repeated because under the hood the same operations will be carried out. However, with large tables, as the number of tuples changes the number and order of the operations may change to improve efficiency, or data that has been previously cached may be accessed first. These may affect the order in which tuples are returned.

As discussed in Chapter 1, having unique tuples in a relation is essential if we are going to be able to identify our data correctly. If, in the relation in Figure A2-2, we found we had another person called John Smith born on 2/6/1988 we would be in trouble, because we would not be able to distinguish the tuples for the two people. We need enough information stored about people so that they can be differentiated. The concept of a primary key (a set of attributes that must be unique for every tuple) ensures uniqueness.

Once we think of our data as sets of tuples then all the power of set operations is at our disposal.

SQL, Algebra, and Calculus

SQL is a language that is mostly based on relational calculus. Relational calculus describes the conditions the retrieved tuples must obey. In the following SQL query, the WHERE clause describes the resulting tuples:

```
SELECT LastName, FirstName, Handicap, PracticeNight
FROM Member, Team
WHERE TeamName = Team AND Handicap < 15;
```

Although SQL is a calculus-based language, more and more keywords suggesting set operations from relational algebra have been included in the syntax over the years. In many cases this makes the queries easier to understand. The preceding query can also be written using the syntax associated with the relational algebra inner join operation, as follows:

```
SELECT LastName, FirstName, Handicap, PracticeNight
FROM Member INNER JOIN Team ON TeamName = Team
WHERE Handicap < 15;
```

The preceding SQL appears to suggest that the join is carried out first and then those tuples with Handicap < 15 are retrieved. This is not the case in practice. SQL is simply a description of the resulting tuples and does not imply how the query will be carried out. The database's query optimizer will determine how the tuples are retrieved, and a good optimizer would carry out the two queries in the same (most efficient) way.

In the remainder of this appendix we will look at a more formal notation for relational algebra and calculus. I will often provide an equivalent SQL expression and will choose one that is similar to the algebra or calculus depending on the section. The important thing to remember is that all SQL expressions are descriptions of the query output, and the way they are expressed does not necessarily determine the operations involved in retrieving the resulting data.

Relational Algebra: Specifying the Operations

With relational algebra, we describe queries by considering a sequence of operations or manipulations on the relations in the database. Some operations act on one relation (*unary* operations), while others are different ways of combining data from two relations (*binary* operations). Every time we perform an operation on one or more relations the result is another relation. This is a very powerful concept and means we can build up complicated queries in small steps by taking the result of one operation and applying another operation to it.

The relational algebra operations and the symbols commonly used to represent them are shown in Table A2-2.

Table A2-2. *Relational Operators and Their Symbols*

Operation	Symbol
select	σ
project	π
Cartesian product	×
union	∪
difference	-
inner join	⋈
intersection	∩
division	÷

The operations are not completely independent. For example, later we will see that an inner join is defined as a Cartesian product followed by a select and a project. The first five operators in Table A2-2 can be used to define the final three, which is why SQL does not need to provide keywords representing division and intersection. However, it is convenient to be able to specify the equivalent SQL for an operation such as inner join because it occurs so frequently in database queries. We will now introduce a more formal notation for each of the operations and show how it can be used to specify queries.

Select

The *select* operation returns just those tuples from a relation that satisfy a particular condition involving the attributes. An example of using a select operation would be to retrieve all the senior members from our Member relation. The Greek letter sigma (σ) stands for the select operation, and the condition, MemberType = 'Senior', is specified in a subscript. The following expression shows the notation for using select to return senior members:

$$\sigma_{\text{MemberType}='Senior'}\left(\text{Member}\right)$$

Each tuple in the relation Member is investigated, and if the tuple meets the condition it is included in the resulting relation. In table terms, the select operator retrieves a subset of the rows of the table. All of the attributes or columns are returned.

In SQL the WHERE clause contains the condition for the select operator and controls the tuples or rows that are returned. The SQL equivalent of the select operation $\sigma_{\text{MemberType}='Senior'}\left(\text{Member}\right)$ is:

```
SELECT *
FROM Member m
WHERE m.MemberType = 'Senior';
```

Note that the SELECT keyword in SQL has nothing directly to do with the relational algebra select operation. More about that in the next section.

Project

The *project* operation returns a relation where the attributes are a subset of the attributes of a relation. The project operator is denoted by π (pi), and the attributes are listed in a subscript. In table terms, project returns a subset of the columns of a table. The following statement would return the FirstName and LastName attributes from every tuple in the relation Member:

$$\pi_{\text{FirstName,LastName}} \left(\text{Member} \right)$$

How many tuples or rows would you expect to be returned from the Member relation as a result of the operation $\pi_{\text{FirstName}}$(Member)? The tuples consist of the single attribute FirstName. The Member relation has 20 tuples, but that includes two occurrences of William, two of Robert, and three of Thomas. Earlier I mentioned that the result of every operation results in another relation. The result of $\pi_{\text{FirstName}}$(Member) must be a set of unique tuples. The duplicates will all be removed, leaving us with 16 unique names.

Think of the project operation as returning all the *unique* combinations of values for the specified attributes.

In SQL the attributes to be returned by the project operator are specified in the SELECT clause. I know this seems perverse, but remember that SQL syntax is based on relational calculus, not on algebra. The SQL equivalent of the project operation $\pi_{\text{FirstName, LastName}}$(Member) is:

```
SELECT DISTINCT FirstName, LastName
FROM Member;
```

Combining Select and Project

Because the result of an algebra operation on a relation always results in another relation, we can apply the operations successively. The following expression first uses the select operation to find all the tuples for senior members (the inner parentheses) and then applies the project operation to return just the names:

$$\pi_{\text{FirstName,LastName}} \left(\sigma_{\text{MemberType='Senior'}} \left(\text{Member} \right) \right)$$

Does the order of the operations make a difference? Consider the following expression where the order of the select and project operations is reversed:

$$\sigma_{\text{MemberType='Senior'}} \left(\pi_{\text{FirstName,LastName}} \left(\text{Member} \right) \right)$$

The tuples resulting from the initial project operation (inner parentheses) have just the two attributes FirstName and LastName. The MemberType attribute is no longer in the tuples, so we cannot use it in the select condition. The algebra expression is not valid.

The SQL statement equivalent to our combined select and project operations is:

```
SELECT FirstName, LastName FROM Member
WHERE MemberType = 'Senior';
```

Because SQL is based on relational calculus rather than algebra, there is no concept of operations or order in the preceding statement. It is just a description of the tuples to be retrieved.

For more complex queries it is sometimes helpful to introduce intermediate relations so we can break up the query into smaller steps. For example, we might call the relation resulting from the select operation SenMemb, as in the following:

$$SenMemb \leftarrow \sigma_{MemberType = 'Senior'} (Member)$$

Now we can use the project operation on the newly named relation SenMemb to return the names:

$$\pi_{FirstName, LastName} (SenMemb)$$

In SQL we can use views to break down queries into simpler steps. A view can be thought of as instructions for creating a new temporary relation:

```
CREATE VIEW SenMemb AS
SELECT * FROM Member
WHERE MemberType = 'Senior';
```

The view can then be used in other queries:

```
SELECT LastName, FirstName
FROM SenMemb;
```

Cartesian Product

The select and project operations are both unary operations, which means they act on a single relation. We will now look at binary operations, which act on two relations. The result of both unary and binary operations is a single relation.

A *Cartesian product* is the most versatile binary operation because it can be applied to any two relations. The notation for a Cartesian product between two relations Member and Team is:

Member × Team

Each tuple in a Cartesian product will have a value for each attribute from the two contributing relations. The tuples in the resulting relation consist of every combination of tuples from the original relations. If one relation has N tuples and the other M, then the resulting relations will have N x M tuples. In table terms, the Cartesian product takes two tables of any shape and produces a table with a column for each column in the original tables and a row for every combination of the original rows. Figure A2-3 shows abbreviated Member and Team tables and their Cartesian product.

MemberID ▾	LastName ▾	FirstName ▾	Team ▾
286	Pollard	Robert	TeamB
461	Reed	Robert	TeamA

Member

TeamName ▾	PracticeNight ▾
TeamA	Tuesday
TeamB	Monday

Team

MemberID ▾	LastName ▾	FirstName ▾	Team ▾	TeamName ▾	PracticeNight ▾
286	Pollard	Robert	TeamB	TeamB	Monday
286	Pollard	Robert	TeamB	TeamA	Tuesday
461	Reed	Robert	TeamA	TeamB	Monday
461	Reed	Robert	TeamA	TeamA	Tuesday

Member X Team

Figure A2-3. *The Cartesian product of Member and Team*

The SQL for a Cartesian product use the keyword CROSS JOIN, as in the following statement:

```
SELECT * FROM Member CROSS JOIN Team;
```

Inner Join

In relational algebra an *inner join* is defined as a Cartesian product followed by a select operation that compares the values of attributes from the two original relations. The attributes being compared must have the same domains.

Referring to the tables in Figure A2-3, we can specify a Cartesian product followed by a select that will return only those tuples where the value of Team is the same as the value of TeamName:

$$\sigma_{\text{Team=TeamName}}\left(\text{Member} \times \text{Team}\right)$$

We can use the join operation to produce an equivalent expression. The join symbol ⋈ is used, and the select, or join, condition is expressed in a subscript as shown in the following expression:

$$\text{Member} \bowtie_{\text{Team=TeamName}} \text{Team}$$

The preceding expressions are *equijoins* where the select condition uses equality. This is the most common type of join. The more general case is a θ-join (theta-join) where the expression can include comparisons such as > and <. A natural join is one where the two relations each have one or more attributes with the same name. By default, the join condition will be equality on the values of the attribute with the same name, and one of those duplicate attributes will be removed from the final result with a project operation.

When we have expressions involving several operations we often have a choice as to the order in which the operations are applied. For example, if we want to retrieve the practice night for Mr. Pollard, we can either select Pollard from the Member relation before the join or afterward from the result of the join. These two options are shown here:

$$\pi_{\text{PracticeNight}}\left(\sigma_{\text{LastName ='Pollard'}}\left(\text{Member} \bowtie_{\text{Team=TeamName}} \text{Team}\right)\right)$$

$$\pi_{\text{PracticeNight}}\left(\left(\sigma_{\text{LastName='Pollard'}} \text{Member}\right) \bowtie_{\text{Team=TeamName}} \text{Team}\right)$$

The tuples resulting from the preceding two expressions are the same; however, the method for obtaining them is quite different. The first will involve first creating a large relation that is the Cartesian product of Member and Team. In the second expression, we reduce the number of tuples in the Member relation to just those for Pollard and then construct a much smaller Cartesian product. Clearly the second expression will be more efficient.

SQL being based on calculus rather than algebra does not imply any ordering of operations. While the SQL statement that follows might suggest that the join is carried out first, it is just a statement describing the tuples to be retrieved:

```
SELECT *
FROM Member INNER JOIN TEAM ON Team = TeamName
WHERE LastName = 'Pollard';
```

The query optimizer in a database system will determine an effective method for carrying out the query.

Union, Difference, and Intersection

Because a relation is defined as a set of tuples, the three binary set operations *union* ($\cup$), *difference* ($-$), and *intersection* ($\cap$) can be used for retrieving information. For relational algebra there is the additional constraint that the two relations involved in these operations must be *union compatible*. This means that the two relations must have the same number of attributes, and the corresponding attributes in each relation must be defined on the same domains.

For example, consider two relations with the following attributes:

```
Staff:{FamilyName, FirstName, Salary}
Students:{LastName, Name, Address, Course}
```

The set operations will help us to retrieve the names of all the people (union), the names of those people who are both students and staff members (intersection), and those who are students but not staff and vice versa (difference). (This, of course, makes naïve assumptions about the uniqueness of names!)

We cannot compare tuples in the relations as they stand because they have different attributes. Staff and Student are not union compatible. One has a Salary while the other has an Address and a Course. However, the names can be compared, as they have the same domains (text) in each relation. We can retrieve just the names by applying a project operation to each of the original relations as follows:

$$\pi_{\text{FamilyName},\text{FirstName}}\left(\text{Staff}\right)$$

$$\pi_{\text{LastName},\text{Name}}\left(\text{Student}\right)$$

Strictly speaking, for union compatibility the attributes should be identical (same name and domain). However, in practice just the domains need to be the same, and the order of the attributes determines what is compared. We can now apply any of the three set operations to the new union-compatible relations. For example:

$$\pi_{\text{FamilyName},\text{FirstName}}\left(\text{Staff}\right)\cup\pi_{\text{LastName},\text{Name}}\left(\text{Student}\right)$$

The SQL expression is:

```
SELECT FamilyName, FirstName FROM Staff
UNION
SELECT LastName, Name FROM Student;
```

We can continue applying operations to the results of our expressions. If taken slowly it is quite straightforward. For example, if we want to find the names and salaries of those staff who are also students we can build up a series of relational algebra operations starting with the initial relations. See the following:

1. Project out the names to get union-compatible relations.

2. Use the intersection operation to find those staff who are also students.

3. Join the result with the Staff relation so we have access to the Salary attribute.

4. Project the Names and Salary attributes.

You can see each of these operations in the following expression–just read the brackets from the inside to the outside:

$$\pi_{\text{FamilyName,FirstName,Salary}} \left(\left(\pi_{\text{FamilyName,FirstName}} \left(\text{Staff} \right) \cap \right. \right.$$

$$\left. \left. \pi_{\text{LastName,Name}} \left(\text{Student} \right) \right) \bowtie_{\text{FamilyName=LastName AND FirstName=Name}} \left(\text{Staff} \right) \right)$$

Union, difference, and intersection are not independent. An intersection can be expressed in terms of two difference operations. Assuming StaffNames and StudentNames are two union-compatible relations, we have that:

$$\text{StaffNames} \cap \text{StudentNames} = \text{StaffNames} - \left(\text{StaffNames} - \text{StudentNames} \right)$$

Draw yourself a sequence of pictures to convince yourself of this.

Some versions of the SQL language do not implement the INTERSECT keyword because the query can be restated, as just seen, using EXCEPT (the SQL syntax for difference). The following SQL query uses the INTERSECT keyword:

```
SELECT * FROM StaffNames
INTERSECT
SELECT * FROM StudentNames;
```

An equivalent query can be constructed using the EXCEPT keyword:

```
SELECT * FROM StaffNames
EXCEPT
(
    SELECT * FROM StaffNames
    EXCEPT
    SELECT * FROM StudentNames
);
```

Division

Division is the last of the relational algebra operations we will consider. The easiest way to understand the division operation is with an example.

If we want to know which members of our club have entered *every* tournament, we need two pieces of information. We need information about the members and the tournaments they have entered, which we can get from the Entry table, and we also need a list of all the tournaments, which comes from the Tournament table.

In Figure A2-4, you can see how division works. It shows the MemberID and TourID attributes from the Entry relation, and the TourID attribute from the Tournament relation. The result of the division is the set of MemberID values that have a tuple in the Entry relation for every value TourID.

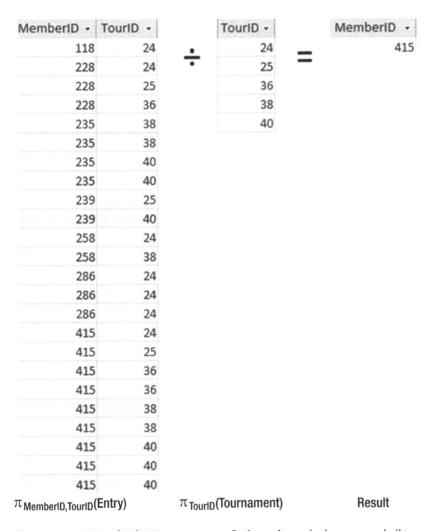

MemberID ▾	TourID ▾
118	24
228	24
228	25
228	36
235	38
235	38
235	40
235	40
239	25
239	40
258	24
258	38
286	24
286	24
286	24
415	24
415	25
415	36
415	36
415	38
415	38
415	40
415	40
415	40

TourID ▾
24
25
36
38
40

MemberID ▾
415

$\pi_{\text{MemberID,TourID}}(\text{Entry})$ ÷ $\pi_{\text{TourID}}(\text{Tournament})$ = Result

Figure A2-4. *Using the division operator to find members who have entered all tournaments*

The relational algebra expression for the division operation in Figure A2-4 is as follows:

$$\pi_{\text{MemberID,TourID}}\left(\text{Entry}\right) \div \pi_{\text{TourID}}\left(\text{Tournament}\right)$$

There is no SQL keyword for the division operator. However, it is possible to express division in terms of other algebraic operations. It can be a bit daunting if presented in one step, so we will take it slowly.

First, find all the members who have entered a tournament and, by way of a Cartesian product, create tuples for each of those members paired with every tournament. We'll call the resulting relation `AllPairs`:

$$\text{AllPairs} = \pi_{\text{MemberID}}\left(\text{Entry}\right) \times \pi_{\text{TourID}}\left(\text{Tournament}\right)$$

Now we will remove from `AllPairs` the pairings that are in the `Entry` table by using a difference operation. If we project out the `MemberID` from the result we will have the IDs for members who are not associated with every tournament.

$$\text{Unmatched} = \pi_{\text{MemberID}}\left(\text{Allpairs} - \pi_{\text{MemberID,TourID}}\left(\text{Entry}\right)\right)$$

By removing these unmatched `MemberID`s from the `MemberID`s in the `Entry` relation we will arrive at the result we require:

$$\text{ResultDivision} = \pi_{\text{MemberID}}\left(\text{Entry}\right) - \text{Unmatched}$$

We can use SQL views to express these same steps in a manageable way. First, create a view with all the pairs of members and tournaments:

```
CREATE VIEW AllPairs AS
SELECT M.MemberID, T.TourID FROM
(SELECT MemberID FROM Entry)M
CROSS JOIN
(SELECT TourID FROM Tournament)T;
```

Now create a view to find the unmatched pairs:

```
CREATE VIEW Unmatched AS
SELECT * FROM AllPairs
EXCEPT
SELECT MemberID,TourID
FROM Entry;
```

Now use these two views to find the result of the division; i.e., the `MemberID` of members who have entered every tournament:

```
SELECT MemberID FROM Entry
EXCEPT
SELECT MemberID FROM Unmatched;
```

If you are brave you could try to combine all these steps into one SQL query; however, we will look at a more manageable way to express the equivalent query using relational calculus in the section on the universal quantifier later in this Appendix.

Relational Calculus: Specifying the Outcome

Relational algebra lets us specify a sequence of operations that eventually results in a set of tuples with the required information. Rather than specifying *how* to do the query, relational calculus describes *what* conditions the resulting data should satisfy. This section provides a very brief introduction to the notation for describing calculus queries without delving into the mathematics.

Simple Calculus Expressions

In informal language, a relational calculus description of a query has the following form:

> *I want the tuples that obey the following conditions....*

More formally we can express the above as:

```
{ m | condition(m) }
```

The part on the left of the bar | specifies the attributes in the tuples we want returned, while the part on the right (often referred to as the *predicate*) describes the criteria they must satisfy. m is called a *tuple variable* and condition(m) is a function that must be true for each of the tuples m returned. Strictly speaking, the preceding notation is called *tuple calculus*. Another equivalent notation, which we will not pursue here, is *domain calculus*.

The following expression means that each returned tuple m must be from the Member relation and must have 'Senior' as the value of the attribute MemberType:

```
{m | Member(m) AND m.MemberType = 'Senior'}
```

We can further refine the expression to specify which attributes of the tuple m should be included in the result:

```
{m.LastName, m.FirstName | Member(m) AND m.MemberType = 'Senior'}
```

Because SQL is based on relational calculus, the equivalent SQL statement is an almost direct translation of the calculus expression, as we see here:

```
SELECT m.LastName, m.FirstName
FROM Member m
WHERE m.MemberType = 'Senior';
```

The part on the left of the bar | in the calculus expression becomes the SELECT clause, Member(m) becomes the FROM clause, and the rest of the expression makes up the WHERE clause. In SQL the m is can be referred to as a table alias, but it is useful to think of it as tuple variable as well.

Free and Bound Variables

The following calculus expression retrieves members' names and the fee associated with their membership type. It is essentially an inner join between the Member and Type relations.

```
{m.LastName, m.FirstName, t.Fee | Member(m) Type(t) AND m.MemberType = t.Type}
```

The tuple variables m and t are referred to as *free variables*. The conditions on the right of the expression cannot be evaluated until we give m and t values. It is usual to refer to the free variables as *ranging* over every tuple in their respective relations and then evaluating the conditions for each combination to see if it should be included in the result. In the body of the book I suggested thinking of the variables as being attached to fingers that move through their respective relations so we can determine if the condition (in this case m.MemberType = t.Type) evaluates to true. Free variables denote the tuples being returned and should always appear on the left of the bar.

Suppose we want to find the names of those members who have entered any tournament. In order to include a member in the result there has to be a tuple for that member in the Entry relation. The symbol ∃, meaning "there exists," is used in the following calculus expression to return the names of those members where a tuple exists in the Entry relation with their MemberID:

```
{m.LastName, m.FirstName | Member(m) AND
∃(e)(
    Entry(e) AND m.MemberID = e.MemberID
    )
}
```

I've spread the expression out on different lines so that the conditions on the variable e are clear. The variable e is referred to as a *bound variable*. It does not appear on the left of the equation and is only used to determine whether the condition on the right side of the expression is true. The free variables (which always appear on the left side of the expression) are the ones for which we consider every possibility. In the preceding expression our free variable m is given the value of every tuple in the Member relation in turn. For each value of m we use the bound variable e to help determine if there is an appropriate tuple in the Entry relation.

Bound variables need to have what is called a *quantifier*, which explains how the variable will be used in calculating the condition statement. In this case we use the *existential quantifier* (∃), which requires us to find a single tuple in the associated relation that satisfies the condition. There is also a *universal quantifier* (∀) that requires *every* tuple in the associated relation to satisfy the condition. We will now look at examples of using these quantifiers and the equivalent SQL statements.

Existential Quantifier and SQL

An expression such as ∃(e)(Entry(e) AND (condition(e)) is true, if we can find a tuple e in the relation Entry that satisfies the specified condition. Let's look at how this query can be represented in SQL:

```
{m.LastName, m.FirstName | Member(m) AND
∃(e)(
    Entry(e) AND e.MemberID = m.MemberID
    )
}
```

First, we'll consider an SQL expression that follows the calculus as closely as possible by using an EXISTS clause:

```
SELECT m.LastName, m.FirstName
FROM Member m
WHERE EXISTS (
    Select * FROM Entry e WHERE e.MemberID = m.MemberID
);
```

As you can see, the SQL is almost a direct translation of the calculus statement. Equivalently, we can represent the existence requirement with a nested query and IN clause:

```
SELECT m.LastName, m.FirstName
FROM Member m
WHERE m.MemberID IN (
    SELECT MemberID FROM Entry e
);
```

Each of preceding SQL statements returns the correct result, but I'm sure you are thinking that they are a complicated way of getting there. The query to find members who have entered a tournament can be more simply expressed as a join between the two relations:

```
SELECT m.LastName, m.FirstName
FROM Member m, Entry e
WHERE e.MemberID = m.MemberID;
```

The preceding SQL statement is not strictly equivalent to the first two. The latter one will return duplicate names, one for each of the tournaments the member has entered. If we look at the first two SQL queries, we see that they are checking each tuple in the Member table (just the once) and looking for a corresponding tuple in the Entry table. The final query considers all combinations of the tuples in Member and Entry and returns any that satisfy the condition (thereby returning the duplicates).

Even though we can remove the duplicates from the final SQL query by adding a DISTINCT keyword, it is considering a different set of tuples for inclusion in the result, and so is responding to a subtly different question than are the two earlier SQL statements. The relational calculus query is very precise, and it is that precision that can be helpful in some situations.

To find members who have not entered a tournament we simply replace ∃ with NOT ∃ (or ∄) in the query to find members who have entered a tournament:

```
{m.LastName, m.FirstName | Member(m) AND
NOT ∃(e)(
    Entry(e) AND m.MemberID = e.MemberID
    )
}
```

The equivalent SQL statement simply requires the addition of the keyword NOT, as in the example here:

```
SELECT m.LastName, m.FirstName
FROM Member m
WHERE NOT EXISTS (
    SELECT * FROM Entry e
    WHERE e.MemberID = m.MemberID
);
```

Universal Quantifier and SQL

The *universal quantifier* ∀ allows us to check that a condition holds for *all* tuples in some set. This is what we require in order for a query to find the names of members who have entered all tournaments. We have looked at this query many times! The relational calculus statement that follows is a straightforward way to express the query:

```
{m.LastName, m.FirstName | Member(m) AND
∀(t)(
    Tournament(t)
    ∃(e)(
        Entry(e) AND
        e.MemberID = m.MemberID
        AND e.TourID = t.TourID
        )
    )
}
```

The calculus statement should be interpreted as "Retrieve the LastName and FirstName for a particular tuple m in Member if for every tuple t in Tournament there exists a tuple e in Entry for the member m and the tournament t."

You will recognize the outcome of this query as the equivalent of the relational algebra division operator. You will also remember that SQL does not have a keyword for division. Sadly, it doesn't have a keyword for the universal quantifier either. Relational calculus can help us out here with the use of the following identity:

```
∀(t)(condition (t)) ≡ NOT ∃(t)(NOT condition(t))
```

This statement means that if we say "for every tuple t a condition holds" then that is the same as saying "there is *no* tuple t for which the condition does *not* hold." We can use this identity to recast our original calculus expression to the following:

```
{m.LastName, m.FirstName | Member(m) AND
NOT ∃(t)(
    Tournament(t)(
    NOT ∃(e)(
        Entry(e) AND e.MemberID = m.MemberID
        AND e.TourID = t.TourID
        )
    )
}
```

Essentially, this says that there is no tuple t in Tournament for which there is not a corresponding tuple e in Entry. This translates quite easily to the SQL statement seen here:

```
SELECT m.LastName, m.FirstName
FROM Member m
WHERE NOT EXISTS (
    SELECT * FROM Tournament t
    WHERE NOT EXISTS (
        SELECT * FROM Entry e
```

```
        WHERE e.MemberID = m.MemberID
        AND e.TourID = t.TourID
    )
);
```

An Example

Let's look at the algebra and calculus for a query that could be a little tricky. We want to find the names of the women who have never played in the Leeston tournament.

Algebra

First, we need to retrieve all entries for the Leeston tournament by joining the `Tournament` and `Entry` tables and then using a select operation:

$$\text{LeestonEntries} \leftarrow \sigma_{\text{TourName ='Leeston'}}(\text{Entry} \bowtie_{\text{TourID=TourID}} \text{Tournament})$$

The words "have never" suggests we need a difference operator. So, we need to find the set of all women by using a select operation on the `Member` table, and then we need to remove the set of people who *have* played at Leeston. In order to use our difference operator we need to have union-compatible relations, so we will project just the `MemberID` from the two sets just described. The following expression will return the IDs of the women who have not entered the Leeston tournament:

$$\text{NonLeestonLadies} \leftarrow \pi_{\text{MemberID}} \left(\sigma_{\text{Gender='F'}} (\text{Member}) \right) - \pi_{\text{MemberID}} (\text{LeestonEntries})$$

Now we need to join `NonLeestonLadies` to the `Member` table so we have access to their names. We can retrieve the final set of names with:

$$\text{Result} \leftarrow \pi_{\text{FirstName, LastName}} \left(\text{Member} \bowtie_{\text{MemberID=MemberID}} \text{NonLeestonLadies} \right)$$

We can now construct an SQL statement that reflects the algebra expression. In the following SQL the most indented rows represent the `LeestonEntries`, the next indentation represents the `NonLeestonLadies` (and has been given that alias), and the outer rows represent the final join and project:

```
SELECT m2.LastName, m2.FirstName FROM
    (SELECT m.MemberID FROM Member m
    WHERE m.Gender = 'F'
    EXCEPT
        SELECT e.MemberID
        FROM entry e INNER JOIN tournament t ON e.tourID = t.tourID
        WHERE t.TourName = 'Leeston'
    )NonLeestonLadies
INNER JOIN Member m2 ON m2.MemberID  = NonLeestonLadies.MemberID;
```

Calculus

Let's approach the same query (the names of the women who have never played in the Leeston tournament) using calculus. I always need to visualize the tuple variables as fingers to get myself started. Figure A2-5 shows the relations that we will need for the query.

MemberID ↓	LastName ↓	FirstName ↓	Gender ▾
118	McKenzie	Melissa	F
138	Stone	Michael	M
153	Nolan	Brenda	F
176	Branch	Helen	F
178	Beck	Sarah	F
228	Burton	Sandra	F
235	Cooper	William	M
239	Spence	Thomas	M
258	Olson	Barbara	F
286	Pollard	Robert	M
290	Sexton	Thomas	M
323	Wilcox	Daniel	M
331	Schmidt	Thomas	M
332	Bridges	Deborah	F
339	Young	Betty	F
414	Gilmore	Jane	F
415	Taylor	William	M
461	Reed	Robert	M
469	Willis	Carolyn	F
487	Kent	Susan	F

Member

MemberID ▾	TourID ▾
118	24
228	24
228	25
228	36
235	38
235	38
235	40
235	40
239	25
239	40
258	24
258	38
286	24
286	24
286	24
415	24
415	25
415	36
415	36

Entry

TourID ▾	TourName ▾
24	Leeston
25	Kaiapoi
36	WestCoast
38	Canterbury
40	Otago

Tournament

Figure A2-5. *Tuple variables required for the query*

We want to retrieve the names of women from the Member table, so we need to consider each tuple in turn. That means m will be our free variable. For each tuple m we need to check that the value of Gender is F and that there is no tuple e in the Entry table that has the same MemberID as m and also has TourID = 24 (the Leeston tournament). Figure A2-5 shows us that although Barbara Olson is a female, we will not include her as she has an entry for the Leeston tournament.

The following calculus expression will retrieve the names of members satisfying the conditions we have just described:

```
{m.LastName, m.FirstName | Member(m) AND m.Gender = 'F'
NOT ∃(e)(
    Entry(e)(
    e.MemberID = m.MemberID
    AND ∃(t)(
        t.TourID = e.TourID
        AND t.Tourname = 'Leeston'
    )
)
}
```

The calculus expression translates directly to the following SQL statement:

```
SELECT m.LastName, m.FirstName
FROM Member m
WHERE m.Gender = 'F'
AND NOT EXISTS (
    SELECT *
    FROM Entry e
    WHERE e.MemberID = m.MemberID
    AND EXISTS (
        Select * FROM Tournament t
        WHERE
        t.TourID = e.TourID
        AND t.Tourname = 'Leeston'
    )
);
```

Conclusion

Having applied a calculus and algebra approach to our query to find women who have not entered a Leeston tournament, we have arrived at two equivalent but quite different-looking SQL queries. There are, no doubt, several other equivalent SQL statements. Testing these in SQL Server 2012 shows that the optimizer produces slightly different execution plans, with the calculus query coming out slightly faster – although adding some indexes could completely change that.

The message from this book is that there are two equivalent but quite different methods of approaching any query. This appendix adds concise notations to help you represent those approaches.

Index

A, B, C

Aggregate operations
    AVG() function, 132
        duplicates, 133
        Nulls, 133
    COUNT() function, 129
        duplicates, 132
        Nulls, 130
    MAX() function, 135
    MIN() function, 135
    nested queries, 144
    ROUND() function, 134
    SUM() function, 135

D

Database design
    combining tables, 204
    foreign key constraints, 199
    inconsistent data, 200
    inconsistent spelling, 202
    numeric values, 201
    primary key, 198
    SQL implementation, 203
Difference, 40–41, 44, 63, 99, 101–102, 112, 118–122, 124, 126–127, 163, 170, 173, 188, 198, 200, 207–208, 217–218, 221–222, 224, 229
Division, 63, 122–126, 128, 134, 142–143, 146, 188, 217, 223–224, 228

E

EXISTS keyword, 57–60, 62, 64

F

Frames, 156–159

G, H

Golf club database, 211–212
Grouping function
    DISTINCT keyword, 143
    HAVING keyword, 140, 142

I

Intersection, 16, 99, 101–102, 107, 111–117, 118, 120, 124, 126–127, 161, 175, 185, 187, 200, 205–207, 210, 217, 221–222

J, K, L, M

Join
    diagrammatic interfaces, 43
    equi-join, 44
    inner join, 35–36, 38, 43, 45–49, 96, 109, 216– 217, 220
    natural, 44
    order, 41
    outcome approach, 36, 42
    outer, 45–49, 73–76, 109–110, 205–206
    process approach
        Cartesian product, 33
        inner join, 35
    techniques
        merge join, 170
        nested loops, 169–170
    SQL Server, 171

N

Nested queries, 51, 53, 58, 60, 62, 65, 115, 144–145, 171–172, 191, 199, 204, 208–209
Normalization, 5, 9

© Clare Churcher 2016
C. Churcher, *Beginning SQL Queries*, DOI 10.1007/978-1-4842-1955-3

Get the eBook for only $5!

Why limit yourself?

Now you can take the weightless companion with you wherever you go and access your content on your PC, phone, tablet, or reader.

Since you've purchased this print book, we're happy to offer you the eBook in all 3 formats for just $5.

Convenient and fully searchable, the PDF version enables you to easily find and copy code—or perform examples by quickly toggling between instructions and applications. The MOBI format is ideal for your Kindle, while the ePUB can be utilized on a variety of mobile devices.

To learn more, go to www.apress.com/companion or contact support@apress.com.

Printed in the United States
By Bookmasters